Quantum Information,
Computation and Communication

Quantum physics allows entirely new forms of computation and cryptography, which could perform tasks currently impossible on classical devices, leading to an explosion of new algorithms, communications protocols, and suggestions for physical implementations of all these ideas. As a result, quantum information has made the transition from an exotic research topic to part of mainstream undergraduate courses in physics.

Based on years of teaching experience, this textbook builds from simple fundamental concepts to cover the essentials of the field. Aimed at physics undergraduate students with a basic background in quantum mechanics, it guides readers through theory and experiment, introducing all the central concepts without getting caught up in details. Worked examples and exercises make this useful as a self-study text for those who want a brief introduction before starting on more advanced books. Solutions are available online at www.cambridge.org/9781107014466.

Jonathan A. Jones is a Professor of Physics at the University of Oxford, where he lectures on quantum information. His main research interest is in NMR implementations of quantum information processing.

Dieter Jaksch is a Professor of Physics at the University of Oxford, where he lectures on quantum information. His main research interest is the theory of ultracold atomic gases, with a focus on their potential applications in quantum information processing.

Quantum Information, Computation and Communication

JONATHAN A. JONES AND DIETER JAKSCH

University of Oxford

CAMBRIDGE
UNIVERSITY PRESS

CAMBRIDGE
UNIVERSITY PRESS

University Printing House, Cambridge CB2 8BS, United Kingdom

One Liberty Plaza, 20th Floor, New York, NY 10006, USA

477 Williamstown Road, Port Melbourne, VIC 3207, Australia

314-321, 3rd Floor, Plot 3, Splendor Forum, Jasola District Centre, New Delhi - 110025, India

79 Anson Road, #06-04/06, Singapore 079906

Cambridge University Press is part of the University of Cambridge.

It furthers the University's mission by disseminating knowledge in the pursuit of
education, learning and research at the highest international levels of excellence.

www.cambridge.org
Information on this title: www.cambridge.org/9781107014466

First published 2012

A catalogue record for this publication is available from the British Library

Library of Congress Cataloging in Publication data
Jones, J. A. (Jonathan A.)
Quantum information, computation and communication /
Jonathan A. Jones, Dieter Jaksch.
p.   cm.
ISBN 978-1-107-01446-6 (hardback)
1. Quantum computers – Textbooks.   2. Information theory in physics – Textbooks.
I. Jaksch, Dieter.   II. Title.
QA76.889.J66   2012
004.1 – dc23   2012010538

ISBN 978-1-107-01446-6 Hardback

Additional resources for this publication at www.cambridge.org/9781107014466

Contents

Introduction

Why yet another book on quantum information theory? Like many lecturers we began writing this text because none of the alternatives seemed quite right. This book is aimed squarely at undergraduate physics students who want a brief but reasonably thorough introduction to the exciting ideas of quantum information, including its applications in computation and communication. It is based on a short course we have taught to fourth-year students at Oxford University since 2004; for the most part it only assumes knowledge of elementary quantum mechanics and linear algebra, and so could even be taught to third-year undergraduates. A brief revision guide to quantum mechanics is provided as an appendix, which should cover any minor points that have been missed.

As the title implies the book is structured in three parts, starting with the basics of quantum information and then applying this to quantum computation and quantum communication. Part I is self-contained, but contains only the barest hints of the exciting applications which attract many people to this field and so might prove unsatisfying on its own. Parts II and III draw heavily on Part I, but are largely independent of each other, and it would be perfectly possible to study only one of these two without the other.

As this text is aimed at physics undergraduates, we believe that it is vital to cover experimental techniques, rather than merely presenting quantum information as a series of abstract quantum operations. We have, however, concentrated on the basic ideas underlying each approach, rather than worrying about particular experimental details. Our aim is not to explain how quantum information processing can actually be achieved, but rather to provide the reader with enough of an introduction to start understanding more specialist sources. The choice of implementations described inevitably reflects our own research interests, but is broad enough to provide an introduction to many important fields. We have, however, almost completely neglected implementations with solid state devices.

In the first two parts of the book we have deliberately taken a cheerfully optimistic approach to quantum information, largely treating quantum systems in terms of a highly idealized picture. We do, of course, consider the problem of decoherence, and briefly describe methods that can be used for error correction, but in general we simply assume that these issues can be ignored. This allows us to address some interesting quantum algorithms without becoming bogged down in experimental details or notational complexities. This approach is, however, inappropriate for quantum communication: while quantum computations generally either work or don't work, communication protocols are dominated by considerations of reliability and efficiency. Furthermore, current experimental implementations are often quite inefficient, and cannot simply be treated as ideal. For this reason we begin Part III with a brief primer on information theory, and also take the opportunity to introduce some slightly more sophisticated notations and mathematical techniques.

The experimental sections will make more sense to a reader with some basic familiarity with atomic physics and optics, but this is not essential. The determined theorist could largely neglect the four experimental chapters describing atom and spin implementations, but would probably be happier with a more focused and rigorous text. The chapter on photon techniques is, however, essential reading before tackling Part III, as quantum communication protocols are closely linked to the photon technologies used to implement them.

Extensive references are included at the end of each chapter for readers who wish to take these ideas further; these largely concentrate on textbooks and review papers, rather than the primary literature, but we have also included a number of landmark papers and papers of particular pedagogical interest.

We have provided a range of exercises at the end of each chapter. Most of these should prove fairly straightforward, but worked answers are available. In some cases we do not prove certain key results, but leave these proofs as an exercise for the reader. Quantum information is a field best understood by doing these basic calculations which lead to familiarity with the underlying ideas.

Our text inevitably goes slightly beyond our course as originally taught, but we have endeavored to keep the range of topics covered very similar. We have, however, been tempted to add one major extension, in the form of a chapter on more advanced quantum algorithms. The material in this chapter is somewhat more challenging than the rest of the text, but even so only provides a very basic introduction to this fascinating field.

We are grateful to all our colleagues who have taught us many of the topics explained in this text. Any errors are, of course, entirely our own.

PART I

QUANTUM INFORMATION

1 Quantum bits and quantum gates

Classical information processing is performed using *bits*, which are just two-state systems, with the two states called 0 and 1. By grouping bits together we can represent arbitrary pieces of information, and by manipulating these bits we can perform arbitrary computations. The corresponding basic element used in quantum information is the quantum bit, or *qubit*. This is simply a quantum system with two orthonormal basis states, which we shall call $|0\rangle$ and $|1\rangle$.

There are many possible physical implementations of a qubit, such as spin states of electrons or atomic nuclei, charge states of quantum dots, atomic energy levels, vibrational states of groups of atoms, polarization states of photons, or paths in an interferometer. At this stage the physical implementation is not important: the idea of a qubit is to abstract the discussion away from physical details. Taking the standard approach of quantum information theory, we shall begin by not worrying too much about the properties of these states, or even what their energies are; we shall simply assume that they are eigenstates of the system's Hamiltonian with known eigenvalues (that is, known energies). This approach allows us to concentrate on the fundamental properties of the system, without considering all the tedious details.[1]

We can in principle perform classical information processing on our quantum system by using the two states $|0\rangle$ and $|1\rangle$ as our logical states 0 and 1 and proceeding in the usual fashion, giving rise to the field of *reversible computing*, which will be explored briefly in Part II. This, however, misses the point. A qubit is not confined to these two states, but can be found in arbitrary superposition states. Although it is not immediately obvious what a state like

$$|\psi\rangle = \alpha|0\rangle + \beta|1\rangle, \tag{1.1}$$

where α and β are complex numbers, actually means in information processing terms, it is clear that quantum bits are in some sense more powerful than their classical equivalents. Quantum information processing is, of course, the art of exploiting these superposition states to perform information processing tasks which are impossible for classical systems. Just as the real power of classical information processing requires groups of bits, the real advantages of quantum information processing only become clear in systems with two or more qubits; for simplicity, however, we are confining ourselves to single isolated qubits at the moment.

Throughout this book we will assume that the reader is familiar with elementary quantum mechanics, and in particular with Dirac's notation for writing quantum states as kets, as

---

[1] A more careful approach is necessary when considering how efficiently quantum information protocols can actually be implemented, as will be seen in Part III.

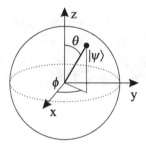

Fig. 1.1   The Bloch sphere: the state $|\psi\rangle$ of a single qubit can be represented as a point on the surface of a sphere with radius 1, or equivalently by the Bloch vector which points from the origin to this point. The state can then be completely described by its co-latitude θ (the angle between the Bloch vector and the z axis) and azimuth angle ϕ (the angle between the projection of the Bloch vector into the xy plane and the x axis).

used above. No particularly sophisticated knowledge is necessary, and if the following two sections make sense then you probably know enough quantum mechanics to understand this book. A brief reminder of some key terms can be found in Appendix A.

1.1 The Bloch sphere

The enormous flexibility of a single qubit in comparison with a classical bit can be most clearly seen using the *Bloch sphere* description of a qubit (Figure 1.1). This also provides a simple but powerful way of visualizing the behavior of a qubit. We begin by looking again at the general state of a single qubit, equation (1.1), and noting that this ket must have unit norm, so that $|\alpha|^2 + |\beta|^2 = 1$. The fact that the state does not change under global phase shifts, so that $e^{i\gamma}|\psi\rangle$ is completely equivalent to $|\psi\rangle$ for *any* real number γ, means that we can always choose α to be *real*, and the normalization constraint is easily imposed by making α and β depend on the cosine and sine of a single parameter. A particularly useful form is to write

$$|\psi\rangle = \cos(\theta/2)|0\rangle + e^{i\phi}\sin(\theta/2)|1\rangle, \tag{1.2}$$

where $0 \leq \theta \leq \pi$ and $0 \leq \phi < 2\pi$. Note that $\theta = 0$ corresponds to $|\psi\rangle = |0\rangle$, and $\theta = \pi$ corresponds to $|\psi\rangle = |1\rangle$; in these extreme cases the value of ϕ is irrelevant.

There is an obvious analogy between the variables θ and ϕ used above and those used in spherical polar coordinates. Clearly any ket $|\psi\rangle$ can be associated with a single point on the surface of a sphere of radius 1 with co-latitude and azimuth angles θ and ϕ; this sphere is usually called the Bloch sphere. Alternatively (and entirely equivalently) a state can be represented as a unit vector (connecting the origin and a point on the Bloch sphere), called the Bloch vector.

Example 1.1 The two basis states $|0\rangle$ and $|1\rangle$, which correspond to the states 0 and 1 of a classical bit, lie at the north and south poles of the Bloch sphere respectively, while a qubit can lie anywhere at all on the surface. Another important group of states is the set of equally weighted superpositions, with $|\alpha| = |\beta| = 1/\sqrt{2}$, which lie on the equator of the Bloch sphere, with the exact position determined by the relative phase of α and β. Unlike $|0\rangle$ and $|1\rangle$, these states have no classical interpretation.

Like any other quantum state, the state of a qubit will evolve under the influence of its Hamiltonian $\mathcal{H}$. The time-dependent Schrödinger equation

$$i\hbar \frac{\partial}{\partial t}|\psi\rangle = \mathcal{H}|\psi\rangle \tag{1.3}$$

has the solution

$$|\psi(t)\rangle = U(t)|\psi(0)\rangle \tag{1.4}$$

with

$$U(t) = \exp(-\mathrm{i}(\mathcal{H}/\hbar)t). \tag{1.5}$$

The evolution of quantum states can also be described using the compact notation

$$|\psi\rangle \xrightarrow{\mathcal{H}t} U|\psi\rangle. \tag{1.6}$$

Since $\mathcal{H}$ is Hermitian, the evolution operator U, usually called the *propagator*, must be unitary.

The discussion above assumes that the Hamiltonian is time-independent, that is it does not vary with time. This will not be true in a quantum computer, which is controlled by varying the Hamiltonian. In many cases, however, the Hamiltonian is *piecewise constant*, that is it has a constant value for some finite length of time, and is then replaced by a different constant value for another finite time period, and so on. In this case the evolution can be described using a series of propagators

$$|\psi\rangle \xrightarrow{\mathcal{H}_1 t_1} \xrightarrow{\mathcal{H}_2 t_2} \xrightarrow{\mathcal{H}_3 t_3} U_3 U_2 U_1 |\psi\rangle \tag{1.7}$$

with $U_1 = \exp[-\mathrm{i}(\mathcal{H}_1/\hbar)t_1]$ and so on. It is, of course, possible to combine the sequence of propagators into a single propagator, $U = U_3 U_2 U_1$. Note that the sequence of Hamiltonians is normally written with time running from left to right (that is the leftmost Hamiltonian is the first to be applied), while the sequence of propagators is always written from right to left, as the rightmost propagator is applied first. The situation is much more complicated when the Hamiltonian varies continuously with time; it is possible to write down a formal solution of the form of equation (1.7), but this is not generally a useful approach. For the moment this issue will simply be ignored.

The fact that any propagator describing the evolution of a quantum system is unitary has several significant consequences. Firstly it means that every propagator has an inverse, and so quantum evolution is *reversible*. (One exception to this general principle is *measurement*, which is discussed in more detail below.) Secondly unitary transformations are *length preserving* and can in general be thought of as *rotations* of the vector describing the quantum state. Since qubits live on the Bloch sphere, the evolution of an isolated qubit under any Hamiltonian corresponds to a rotation of the vectors on the Bloch sphere.

1.2 Density matrices and Pauli matrices

It is frequently convenient to describe the state of a qubit using a vector, written using the basis states $|0\rangle$ and $|1\rangle$ (the computational basis). The basis states take the simple forms

$$|0\rangle = \begin{pmatrix} 1 \\ 0 \end{pmatrix} \quad \text{and} \quad |1\rangle = \begin{pmatrix} 0 \\ 1 \end{pmatrix}. \tag{1.8}$$

(There is a potential ambiguity in any description of quantum bits, as to whether $|0\rangle$ and $|1\rangle$ are defined as shown here, or the other way round; fundamentally, of course, the choice does not matter, as long as one is consistent.) In this basis equation (1.1) can be written as

$$|\psi\rangle = \begin{pmatrix} \alpha \\ \beta \end{pmatrix}, \tag{1.9}$$

while the corresponding bra can be written as

$$\langle\psi| = \begin{pmatrix} \alpha^* & \beta^* \end{pmatrix}, \tag{1.10}$$

as a bra is the adjoint of the corresponding ket, and for a matrix the adjoint is the complex conjugate of the transpose.

Bras and kets are frequently combined by taking the *inner product*, such as

$$\langle\psi|\psi\rangle = \begin{pmatrix} \alpha^* & \beta^* \end{pmatrix} \begin{pmatrix} \alpha \\ \beta \end{pmatrix} = \alpha^*\alpha + \beta^*\beta = 1 \tag{1.11}$$

but they can also be combined using the *outer product*

$$|\psi\rangle\langle\psi| = \begin{pmatrix} \alpha \\ \beta \end{pmatrix} \begin{pmatrix} \alpha^* & \beta^* \end{pmatrix} = \begin{pmatrix} \alpha\alpha^* & \alpha\beta^* \\ \beta\alpha^* & \beta\beta^* \end{pmatrix}. \tag{1.12}$$

This outer product is called a *density matrix* description of the state. As we will see later, density matrices can provide a very useful approach to describe qubits whose states are at least partly unknown, called *mixed states*, but for the moment we will use them simply to explore an alternative to the ket notation for *pure states*.

It is obvious from the form of equation (1.12) that the density matrix describing a qubit is Hermitian, and has trace one; these are in fact general properties which apply to all density matrices. A two-by-two matrix can always be expanded as a weighted sum of four basic matrices (a matrix basis), and perhaps the most useful basis is provided by the Pauli matrices

$$\sigma_0 = \begin{pmatrix} 1 & 0 \\ 0 & 1 \end{pmatrix} \quad \sigma_x = \begin{pmatrix} 0 & 1 \\ 1 & 0 \end{pmatrix} \quad \sigma_y = \begin{pmatrix} 0 & -i \\ i & 0 \end{pmatrix} \quad \sigma_z = \begin{pmatrix} 1 & 0 \\ 0 & -1 \end{pmatrix}, \tag{1.13}$$

where the usual set of three matrices has been extended to include the *identity matrix* $\sigma_0 = \mathbb{1}$. As the Pauli matrices are Hermitian, a density matrix can be written as

$$|\psi\rangle\langle\psi| = \tfrac{1}{2}\left(\sigma_0 + s_x\sigma_x + s_y\sigma_y + s_z\sigma_z\right), \tag{1.14}$$

where s_x, s_y and s_z are three *real* coefficients. This might seem excessive, as we know that any pure state can be described using only two real numbers (θ and ϕ), but it is easily shown that s_x, s_y and s_z are not completely independent, being the three components of a vector of unit length; this vector is identical to the Bloch vector, discussed above.

Qubits can also be found in mixed states, which are just weighted sums of pure states of the form

$$\rho = \sum_n P_n|\psi_n\rangle\langle\psi_n|, \tag{1.15}$$

where $P_n \geq 0$ is the contribution of the pure state $|\psi_n\rangle\langle\psi_n|$ to the mixture (the probability of the pure state occurring in the mixture). Clearly such mixed states are Hermitian, and as the probabilities of the various contributions must sum to one ($\sum_n P_n = 1$), the density matrix must have trace one. It can be shown that any mixed state of a single qubit corresponds to a point *inside* the Bloch sphere.

It is useful to be able to calculate the evolution of states described using a density matrix rather than a ket vector. This problem can be addressed directly by solving the Liouville–von Neumann equation (the density matrix equivalent of the time-dependent Schrödinger equation), but it is simple to proceed by analogy. The evolution of a bra vector is clearly closely related to the evolution of the corresponding ket vector

$$(U|\psi\rangle)^\dagger = \langle\psi|U^\dagger, \tag{1.16}$$

and so the density matrix description of a pure state evolves according to

$$|\psi\rangle\langle\psi| \xrightarrow{\mathcal{H}t} U|\psi\rangle\langle\psi|U^\dagger \tag{1.17}$$

and the linearity of the operations guarantees that a mixed state will evolve in the same way.

We have already noted that the Pauli matrices are Hermitian, and thus provide a natural basis for describing the density matrix corresponding to a qubit. In the same way, the fact that any Hamiltonian is Hermitian means that any Hamiltonian for a single qubit can be written as a weighted sum of the four Pauli matrices, equation (1.13), where the weights are *real*. This means that the Pauli matrices provide a natural language for describing single-qubit Hamiltonians as well as single-qubit states. Furthermore the Pauli matrices are unitary, and so correspond to possible propagators. As we shall see later, the Pauli matrices viewed as propagators correspond to important quantum logic gates. It might seem that using the Pauli matrices to describe quantum states, Hamiltonians, propagators, and logic gates will inevitably lead to confusion, but in practice such problems rarely occur.

The fact that the Pauli matrices are *both* unitary and Hermitian has the interesting consequence that

$$\sigma_\alpha^2 = \sigma_0, \tag{1.18}$$

where σ_α are the usual Pauli matrices, with α equal to x, y or z. This observation can be combined with the series expansion of an exponential operator to show that

$$\exp(-\mathrm{i}\theta\,\sigma_\alpha) = \cos(\theta)\sigma_0 - \mathrm{i}\sin(\theta)\sigma_\alpha \tag{1.19}$$

without diagonalizing any matrices, making it easy to calculate many single-qubit propagators.

Finally we note that the propagator corresponding to a Hamiltonian that is some multiple of σ_0 is simply a global phase shift, which has no physical significance. In essence this occurs because adding multiples of σ_0 corresponds to moving the zero-point of the energy scale, which has no physical significance.

1.3 Quantum logic gates

The basic idea of quantum information processing is that information is stored in quantum bits and processed by quantum logic gates. Just as classical logic gates take classical bits from one state to another, so quantum logic gates take qubits from one state to another. This can be achieved by modifying the system's Hamiltonian, by applying additional *control fields* to the background Hamiltonian which underlies the system.

Applying Hamiltonians will cause qubits to evolve under unitary transformations, which are reversible. With classical bits there are only two reversible gates that act on a single bit: the NOT gate, which takes a bit in state 0 into state 1 and *vice versa*, and the IDENTITY gate, which just leaves the bit unchanged. (It may seem excessive to consider trivial gates such as IDENTITY, but the formalism works better if they are included.) There are also two irreversible gates, SET which sets a bit to 1 whatever its initial state, and CLEAR which sets a bit to 0. Clearly these two cannot be achieved with unitary transformations, and so we will neglect them for the moment.

Returning to the two reversible gates, we must first find unitary propagators that implement them. Clearly σ_0 will perform IDENTITY as

$$\begin{pmatrix} 1 & 0 \\ 0 & 1 \end{pmatrix}\begin{pmatrix} 1 \\ 0 \end{pmatrix} = \begin{pmatrix} 1 \\ 0 \end{pmatrix} \quad \text{and} \quad \begin{pmatrix} 1 & 0 \\ 0 & 1 \end{pmatrix}\begin{pmatrix} 0 \\ 1 \end{pmatrix} = \begin{pmatrix} 0 \\ 1 \end{pmatrix} \tag{1.20}$$

while σ_x corresponds to NOT as

$$\begin{pmatrix} 0 & 1 \\ 1 & 0 \end{pmatrix}\begin{pmatrix} 1 \\ 0 \end{pmatrix} = \begin{pmatrix} 0 \\ 1 \end{pmatrix} \quad \text{and} \quad \begin{pmatrix} 0 & 1 \\ 1 & 0 \end{pmatrix}\begin{pmatrix} 0 \\ 1 \end{pmatrix} = \begin{pmatrix} 1 \\ 0 \end{pmatrix}. \tag{1.21}$$

We now have to find Hamiltonians which can give rise to these propagators. Obviously σ_0 can in principle be achieved simply by doing nothing at all, but in fact the IDENTITY gate is slightly more subtle than it might seem, as the state of the qubit will continue to evolve under the background Hamiltonian even when no additional control fields are applied, and unless the IDENTITY gate is instantaneous this background evolution must be considered. Achieving σ_x is only slightly more difficult: using equation (1.19)

$$\exp(-\mathrm{i}\pi\sigma_x/2) = -\mathrm{i}\sigma_x \tag{1.22}$$

(the reason for dividing the σ_x by 2 will soon become clear). The factor of $-i$ is just a global phase, and so can be ignored. Thus a NOT gate can be achieved by evolving the qubit under a Hamiltonian proportional to σ_x for an appropriate time.

Once again, however, the situation is subtler than it might at first seem. The obvious approach is just to apply a control field which generates a Hamiltonian proportional to σ_x, but this is not quite right as the background Hamiltonian will still be present. A brute force solution is just to make the control field very large in comparison with the background Hamiltonian, but this is rarely practical. A better approach is to apply a weak control field which oscillates at a resonance frequency of the system, enabling the effect of the control field to build up over many cycles. This point will be explored in considerable detail in subsequent chapters.

The quantum NOT gate behaves exactly like a classical NOT gate when applied to basis states, but it can also be applied to more general states. The linearity of quantum mechanics means that logic gates can be trivially extended to deduce their actions on superposition states:

$$\begin{pmatrix} 0 & 1 \\ 1 & 0 \end{pmatrix} \begin{pmatrix} \alpha \\ \beta \end{pmatrix} = \begin{pmatrix} \beta \\ \alpha \end{pmatrix}. \tag{1.23}$$

The effect of this gate can be better understood by considering its effect on the Bloch sphere. Rewriting the general state in polar coordinates as before,

$$\begin{pmatrix} 0 & 1 \\ 1 & 0 \end{pmatrix} \begin{pmatrix} \cos(\theta/2) \\ e^{i\phi} \sin(\theta/2) \end{pmatrix} = \begin{pmatrix} e^{i\phi} \sin(\theta/2) \\ \cos(\theta/2) \end{pmatrix} \tag{1.24}$$

$$= e^{i\phi} \begin{pmatrix} \sin(\theta/2) \\ e^{-i\phi} \cos(\theta/2) \end{pmatrix} \tag{1.25}$$

$$= e^{i\phi} \begin{pmatrix} \cos([\pi - \theta]/2) \\ e^{-i\phi} \sin([\pi - \theta]/2) \end{pmatrix}, \tag{1.26}$$

shows that (neglecting the irrelevant global phase) the effect of a NOT gate is to negate both the latitude and longitude coordinates. A little thought shows that this is equivalent to rotating the Bloch vectors by $180°$ around the x axis. The significance of equation (1.22) should now be clear: the effect of applying some Hamiltonian to a qubit is to rotate the states around an axis *parallel* to the Hamiltonian. The angle of rotation depends on both the intrinsic strength of the Hamiltonian, and the time for which it is applied.

Thinking of the NOT gate as a $180°$ rotation also makes sense when considering the effect of applying two NOT gates in sequence. Clearly this should have no overall effect, and it is comforting to note that the effect of two successive $180°$ rotations is equivalent to a $360°$ rotation, which leaves any state on the Bloch sphere unchanged.[2]

---

[2] In fact careful thought shows that a $360°$ rotation does not leave a state completely unchanged, but applies a phase factor of -1; this is an example of *spinor* behavior. If the phase shift is a global phase then it can be ignored, but in two-qubit gates spinor behavior can be used to generate useful phase shifts as we shall see in Section 10.6. This global phase can be traced to the factor of $-i$ in equation (1.22); note that two successive σ_x operations would leave a state *completely* unchanged. This point is pursued further below.

Reversing this approach we can also think about rotations through smaller angles, such as a 90° rotation around the x axis. This has the propagator

$$\exp[-i\,\pi/2\,\sigma_x/2] = \frac{1}{\sqrt{2}}\begin{pmatrix} 1 & -i \\ -i & 1 \end{pmatrix}, \tag{1.27}$$

which acts to convert basis states into superpositions. Applying this propagator twice gives a NOT gate, and so it is called the SQUARE-ROOT-OF-NOT gate. Clearly this gate has no classical equivalent: it is a purely quantum logic gate.

This is not, of course, the only purely quantum logic gate: there are an infinite number of such gates! In general any rotation on the Bloch sphere (that is, a rotation by any angle around any axis) can be considered as a quantum logic gate, and can be implemented by applying an appropriate Hamiltonian for an appropriate time. For the moment we will briefly consider two of the more important gates: the Hadamard gate and the phase gate.

The Hadamard gate, usually indicated by the letter H, is similar to the SQUARE-ROOT-OF-NOT gate, but with subtly different effects. It is described by the propagator

$$H = \frac{1}{\sqrt{2}}\begin{pmatrix} 1 & 1 \\ 1 & -1 \end{pmatrix} \tag{1.28}$$

and so acts on the basis states to give

$$|0\rangle \xrightarrow{\text{H}} \frac{1}{\sqrt{2}}(|0\rangle + |1\rangle) = |+\rangle \quad \text{and} \quad |1\rangle \xrightarrow{\text{H}} \frac{1}{\sqrt{2}}(|0\rangle - |1\rangle) = |-\rangle. \tag{1.29}$$

The two states $|+\rangle$ and $|-\rangle$, which lie on the equator of the Bloch sphere, play a central role in quantum information processing and will be seen frequently in subsequent chapters.

Unlike the SQUARE-ROOT-OF-NOT gate the Hadamard gate is self-inverse,

$$|+\rangle \xrightarrow{\text{H}} |0\rangle \quad \text{and} \quad |-\rangle \xrightarrow{\text{H}} |1\rangle, \tag{1.30}$$

so that applying it twice is equivalent to doing nothing. This means that the Hadamard gate must correspond to a 180° rotation, and it is in fact equivalent to a 180° rotation around an axis tilted at 45° degrees from the x axis toward the z axis.

The phase gate is usually indicated by the letter S, and can be thought of as the SQUARE-ROOT-OF-σ_z gate. It is described by the propagator

$$S = \begin{pmatrix} 1 & 0 \\ 0 & i \end{pmatrix} \tag{1.31}$$

and its effect is simply to change the phase of $|1\rangle$ by 90°, while leaving $|0\rangle$ unaffected, or, equivalently, to rotate the Bloch sphere by 90° around the z axis. Note that the classical states 0 and 1, which lie at the north and south poles of the Bloch sphere, are not affected by this rotation, but the phase of a superposition *will* be changed. Thus even though S does not interconvert basis states and superposition states it is, like H, a purely quantum gate.

It is possible to describe quantum gates in many different ways, and this gives rise to a range of notations for discussing them. For example the NOT gate can also be written as X, as σ_x, or as π_x or 180°_x. The decision between these is often a matter of context and the background of the person discussing the gate! Researchers with a background in

	Gate	Pauli	Implementation	Matrix form
Table 1.1 Some key quantum logic gates described using different notations (global phases have been neglected)				
	$\mathbb{1}$	σ_0		$\begin{pmatrix} 1 & 0 \\ 0 & 1 \end{pmatrix}$
	X	σ_x	180°_x	$\begin{pmatrix} 0 & 1 \\ 1 & 0 \end{pmatrix}$
	Y	σ_y	180°_y	$\begin{pmatrix} 0 & -i \\ i & 0 \end{pmatrix}$
	Z	σ_z	180°_z	$\begin{pmatrix} 1 & 0 \\ 0 & -1 \end{pmatrix}$
	$S = \sqrt{Z}$		90°_z	$\begin{pmatrix} 1 & 0 \\ 0 & i \end{pmatrix}$
	$T = \sqrt{S}$		45°_z	$\begin{pmatrix} 1 & 0 \\ 0 & e^{i\pi/4} \end{pmatrix}$
	$\sqrt{X}$		90°_x	$\frac{1}{\sqrt{2}} \begin{pmatrix} 1 & -i \\ -i & 1 \end{pmatrix}$
	$\sqrt{Y}$		90°_y	$\frac{1}{\sqrt{2}} \begin{pmatrix} 1 & -1 \\ 1 & 1 \end{pmatrix}$
	H			$\frac{1}{\sqrt{2}} \begin{pmatrix} 1 & 1 \\ 1 & -1 \end{pmatrix}$

computer science would tend to use the most abstract notation, X, while physicists studying the theory of quantum information processing might instead choose the Pauli matrix form, σ_x. By contrast, experimental physicists, who are interested in actually building quantum computers, would usually use the descriptions π_x or 180°_x, as these correspond most closely to a physical process. It is usually a good idea to keep an open mind, and be ready to use whatever notation is around. A list of important gates can be found in Table 1.1.

There is, however, one important distinction between the theoretician's X and σ_x, on the one hand, and π_x or 180°_x on the other, and this is the matter of global phases. It is clear from equation (1.22) that 180°_x is *not* exactly the same as σ_x, but differs by a global phase factor of $-i$. The list of gates in Table 1.1 ignores this point, and the different notations are only equivalent up to global phases. In the single-qubit case this global phase is completely irrelevant, but in systems with two or more qubits it is necessary to be a little more careful.

1.4 Quantum networks

Just as a single bit is not much use on its own, very little can be achieved with a single logic gate. Effective information processing requires that gates be joined together to form *networks*, and the same approach can be used with quantum logic gates. A classical logic network can be built using only one- and two-bit gates (such as AND, OR and NOT), and it can be shown that any quantum logic network can be built out of one-qubit and two-qubit quantum logic gates. Clearly quantum networks will only be really useful when applied to systems with more than one qubit, but even with a single isolated qubit the idea has some use. Gate networks can be used both to explain some classic experiments, such as Ramsey fringes and spin echoes, which will be explored in later chapters, and also to build single-qubit gates out of other gates.

Example 1.2 Show that the Hadamard gate H can be implemented using the sequence of rotations $90_z^{\circ}\, 90_x^{\circ}\, 90_z^{\circ}$.

Solution

As global phases may be neglected, we can use the matrix forms given in Table 1.1. The propagator for this sequence of gates can be obtained by multiplying out the propagators for the individual gates:

$$\begin{pmatrix} 1 & 0 \\ 0 & i \end{pmatrix} \frac{1}{\sqrt{2}} \begin{pmatrix} 1 & -i \\ -i & 1 \end{pmatrix} \begin{pmatrix} 1 & 0 \\ 0 & i \end{pmatrix} = \frac{1}{\sqrt{2}} \begin{pmatrix} 1 & 1 \\ 1 & -1 \end{pmatrix}, \tag{1.32}$$

which is a Hadamard gate.

Example 1.3 Find the overall effect of the network HSSH

$$-\boxed{\text{H}}-\boxed{\text{S}}-\boxed{\text{S}}-\boxed{\text{H}}- \tag{1.33}$$

which corresponds to applying first a Hadamard gate, then a phase gate, then another phase gate, and finally a Hadamard gate.

Solution

The effect of this network could be deduced by applying the gates in sequence to a qubit in a general state, but as before it is more useful to consider the network directly, by simply multiplying out the constituent propagators

$$\frac{1}{\sqrt{2}} \begin{pmatrix} 1 & 1 \\ 1 & -1 \end{pmatrix} \begin{pmatrix} 1 & 0 \\ 0 & i \end{pmatrix} \begin{pmatrix} 1 & 0 \\ 0 & i \end{pmatrix} \frac{1}{\sqrt{2}} \begin{pmatrix} 1 & 1 \\ 1 & -1 \end{pmatrix} = \begin{pmatrix} 0 & 1 \\ 1 & 0 \end{pmatrix} \tag{1.34}$$

to find that this network is equivalent to a NOT gate. Since $SS = \sigma_z$ this network can also be written as $H\sigma_z H = \sigma_x$, or even more simply as $HZH = X$. Other useful identities include $X^2 = Y^2 = Z^2 = H^2 = \mathbb{1}$ and $HXH = Z$.

The network notation does give rise to one serious ambiguity of notation which we have sidestepped in the examples above. When describing a process by a sequence of operators, the operators are applied from right to left, so that the first operator applied is the rightmost operator written in the sequence. By contrast, networks are usually written running from left to right, so that the first operator applied is the leftmost operator written in the network.

In some cases, therefore, it can be unclear whether to apply the gates from right to left or left to right! In the networks above, of course, this distinction is irrelevant as the networks are symmetric, but in general it is necessary to be careful to ensure that this ambiguity does not become a problem.

An important example of building quantum logic gates out of networks is provided by the Hadamard gate. There are many different ways of implementing this, but perhaps the most useful approach is to relate the Hadamard to a $90°$ rotation. We have already considered a $90°_x$ rotation, and a $90°_y$ rotation can be used in much the same way. From the form of the $90°_y$ operator in Table 1.1 it is obvious that it is closely related to a Hadamard, and calculations confirm that the Hadamard is equivalent to a $180°_z$ operation followed by a $90°_y$ operation, or to a $90°_{-y}$ followed by a $180°_z$.

While experimentalists sometimes worry about how best to implement H, theoreticians usually start from the other extreme and seek the smallest possible set of universal single-qubit gates, which allows *all* possible single-qubit gates to be implemented (strictly speaking, to be approximated to arbitrary accuracy) using a network of gates from the basic set. A key result is that the combination of the Hadamard gate and the small-angle phase gate $T = \sqrt{S} = \sqrt[4]{Z}$ is universal.

1.5 Initialization and measurement

So far we have only considered unitary gates (gates that can be described by unitary matrices), but some important gates are obviously not unitary. For example, consider the CLEAR gate

$$|\psi\rangle -\boxed{0}- |0\rangle, \tag{1.35}$$

which sets a qubit to the state $|0\rangle$ whatever its initial state is; clearly this process cannot be described by matrix multiplication. This might seem problematic, as evolution of a quantum system under a Hamiltonian is *always* unitary, and it is not clear how a quantum system can evolve other than in response to a Hamiltonian.

The solution to this quandary is that while a single isolated qubit can only undergo unitary evolution, there isn't really any such thing as an isolated qubit. The fact that we can use control fields to alter the state of the qubit means that the qubit *must* have some interaction with the rest of the world. It can be shown that non-unitary evolutions of a qubit can be achieved by performing a unitary evolution on a composite system, comprising the qubit and some environment, and then ignoring the state of the environment. A detailed analysis of this process clearly requires an understanding of two-qubit systems, and so is beyond the scope of this chapter; for the moment it suffices to note that non-unitary operations can be performed.

Another important non-unitary gate is the READOUT gate, or measurement gate

$$|\psi\rangle -\boxed{\measuredangle}- , \tag{1.36}$$

which simply performs a classical measurement of the state of a single qubit. A full discussion of what measuring the state of a quantum system really means would be very complicated, and even to some extent controversial, but fortunately it is easy to give an accurate mathematical description of what the measurement process *does* to the quantum state. As usual we start by considering a single qubit in a general state

$$|\psi\rangle\langle\psi| = \begin{pmatrix} \alpha \\ \beta \end{pmatrix} \begin{pmatrix} \alpha^* & \beta^* \end{pmatrix} = \begin{pmatrix} \alpha\alpha^* & \alpha\beta^* \\ \beta\alpha^* & \beta\beta^* \end{pmatrix} \tag{1.37}$$

and then consider the state of the qubit *after* the measurement. Assuming we measure in the computational basis, we know that the result of the measurement will be either that the qubit is in state $|0\rangle$, or that it is in state $|1\rangle$, and that after the measurement the qubit will be found in the appropriate state. We also know that the probability of getting the result $|0\rangle$ is given by $|\alpha|^2 = \alpha\alpha^*$, and the probability of getting the result $|1\rangle$ is given by $|\beta|^2 = \beta\beta^*$.

We could choose to stop the discussion here, but it would be useful to be able to describe the state of the system *after* the measurement in the language we have used before. If we know the result of the measurement then this is simple, and the qubit will be in either $|0\rangle$ or $|1\rangle$. Suppose, however, that we wish to consider both possible outcomes together; in this case we don't know what the state of the system is after the measurement, because we don't know what the result of the measurement was! We can, however, make probabilistic statements about it, and this is the way to proceed. Clearly the final state is a mixed state, with the form of equation (1.15), and can be written as

$$\rho = \alpha\alpha^*|0\rangle\langle0| + \beta\beta^*|1\rangle\langle1| \tag{1.38}$$

$$= \alpha\alpha^* \begin{pmatrix} 1 & 0 \\ 0 & 0 \end{pmatrix} + \beta\beta^* \begin{pmatrix} 0 & 0 \\ 0 & 1 \end{pmatrix} \tag{1.39}$$

$$= \begin{pmatrix} \alpha\alpha^* & 0 \\ 0 & \beta\beta^* \end{pmatrix} \tag{1.40}$$

showing that, from a mathematical point of view, the effect of a measurement whose outcome is unknown is simply to force the off-diagonal elements of the density matrix to zero.

This process is sometimes called *decoherence* as the elements of the density matrix which are lost are those that correspond to the system being in a *coherent superposition* state. Decoherence is almost always the enemy of quantum information processing, and great effort is expended attempting to control it. Here we simply note that the random interactions between a quantum system and its environment have the same form as measurements. Thus decoherence results from the environment "measuring" the state of the system, and so the system must be well insulated from the surroundings if it is to exhibit interesting quantum behavior.

We could, of course, choose to measure the qubit using some other basis. Measuring in the computational basis is sometimes called a Z-measurement, as the possible outcomes are the eigenstates of the Z operator; equivalently a qubit is projected onto the z axis of the Bloch sphere. We could instead perform an X-measurement, for which the possible results are

$|+\rangle$ and $|-\rangle$, the eigenstates of X which lie on the $\pm x$ axes of the Bloch sphere.[3] While this would make the process appear more complicated, it does not change any fundamentals: a measurement of a single qubit in any basis can always be achieved by using a measurement in the computational basis preceded and followed by appropriate unitary transformations. Decoherence can arise from environmental "measurements" in any basis, depending on the form of the underlying interactions.

1.6 Experimental methods

In the next three chapters we will consider methods for implementing unitary single-qubit logic gates in three experimental systems: trapped atoms and ions, nuclear and electron spins, and single photons. These three examples have been chosen not just because of their importance in current and possible future implementations of quantum information processing, but also because many other experimental systems are broadly analogous to one of these three. Methods for implementing the non-unitary processes corresponding to initialization and measurement will only be addressed very briefly; more detailed treatments of these, together with methods for implementing multi-qubit logic gates, can be found in Parts II and III.

Further reading

The definitive text on quantum information processing, Nielsen and Chuang (2000), is challenging, but some sections are quite straightforward and it is certainly worth a look. A wide range of intermediate texts is now available, such as Estève *et al.* (2003), Le Bellac (2006), and Stolze and Suter (2008).

Exercises

1.1   Show that if $|\psi\rangle = \cos(\theta/2)|0\rangle + \sin(\theta/2)e^{i\phi}|1\rangle$ then

$$|\psi\rangle\langle\psi| = \tfrac{1}{2}\left(\sigma_0 + s_x\,\sigma_x + s_y\,\sigma_y + s_z\,\sigma_z\right).$$

Show that $\mathbf{s} = (s_x, s_y, s_z)$ (the Bloch vector) has unit length, and so $|\psi\rangle\langle\psi|$ can be represented by a point on the unit sphere (Bloch sphere).

1.2   Show that any mixed state of a single qubit can be written as a point *in* the Bloch sphere. What point does $\tfrac{1}{2}\sigma_0$ correspond to?

---

[3] A more formal approach to measurement in terms of operators will be developed in Part III.

1.3   The maximally mixed state is frequently described as an equal mixture of the two basis states $|0\rangle$ and $|1\rangle$. Show that it can also be described as an equal mixture of the two superposition states $|+\rangle$ and $|-\rangle$.

1.4   Use a Bloch sphere description to explain why there are an infinite number of such decompositions of the maximally mixed state.

1.5   Show that $\sigma_\alpha^2 = \sigma_0$, where σ_α are the usual Pauli matrices, with α equal to x, y or z. Hence use a series expansion to show that $\exp(-i\theta\,\sigma_\alpha/2) = \cos(\theta/2)\sigma_0 - i\sin(\theta/2)\sigma_\alpha$ without diagonalizing any matrices.

1.6   Using matrix propagators show that the Hadamard gate can be implemented as $90_y^\circ\,180_x^\circ$ (where rotations are written from left to right; note that propagators must be applied from right to left). Show that other possible implementations include $180_x^\circ\,90_{-y}^\circ$, $90_{-y}^\circ\,180_z^\circ$, and $180_z^\circ\,90_y^\circ$.

1.7   We have used matrices to show that HZH = X; now show that HXH = Z without multiplying matrices.

1.8   Rewrite the general state of a qubit $|\psi\rangle = \alpha|0\rangle + \beta|1\rangle$ in the X-basis (that is as a superposition of $|+\rangle$ and $|-\rangle$). Show that the result of an X-measurement on this state is identical to the effect of applying a Hadamard gate, performing a Z-measurement, and then applying another Hadamard gate.

1.9   Explain *why* the result of the previous question works, and why any single-qubit measurement gate can be achieved by combining unitary transformations with a Z-measurement.

2 An atom in a laser field

In this chapter we will use a succession of different methods to calculate the interaction between an atom or ion and the light field from a laser. We will see that the effect of the light is to cause *transitions* between different energy levels in the atom, but that these transitions will normally only occur if the frequency of the light is tuned to match the energy gap between the levels

$$h\nu = \hbar\omega = E_f - E_i, \tag{2.1}$$

so that the light is *resonant* with the transitions. All the methods used in this chapter are *semi-classical* treatments, in which we treat the light field as a classical system; a more complete treatment, in which the light field is also quantized, requires methods from *quantum optics*.

Atoms have an infinite number of energy levels, and might seem to be rather complex systems, but the resonance condition means that our treatment of them can be greatly simplified. In most cases it will be sufficient to consider a *two-level atom*, which is assumed to have a ground state $|g\rangle$ and a single excited state $|e\rangle$, and a laser field which is close to resonance with this transition. Other transitions are far from resonance and so can be ignored.

2.1 Time-dependent systems

Consider a quantum mechanical system with a Hamiltonian $\mathcal{H}^0$, which is subjected to a time-varying perturbation $\mathcal{H}^1(t)$. The total Hamiltonian of the system is then

$$\mathcal{H} = \mathcal{H}^0 + \mathcal{H}^1(t). \tag{2.2}$$

As usual the eigenstates of $\mathcal{H}^0$ form a complete set, and so we can write the wavefunction of the system in this basis

$$|\psi(t)\rangle = \sum_j c_j(t)|j\rangle, \tag{2.3}$$

with the time dependence of $|\psi(t)\rangle$ arising from the time dependence of the coefficients. If there was no perturbation present then these coefficients would still oscillate at their natural frequencies,

$$c_j(t) = c_j(0)\mathrm{e}^{-\mathrm{i}E_j t/\hbar}, \tag{2.4}$$

and so it is useful to separate the time variation into that which would occur without the perturbation, and any additional variation which can be ascribed to the perturbation. Thus we write

$$|\psi(t)\rangle = \sum_j d_j(t)\mathrm{e}^{-\mathrm{i}E_j t/\hbar}|j\rangle \tag{2.5}$$

with all the interesting behavior now found in the values of $d_j(t)$. Now we know from the time-dependent Schrödinger equation that $[\mathrm{i}\hbar\,\partial/\partial t - \mathcal{H}^0 - \mathcal{H}^1(t)] = 0$, and applying this operator to equation (2.5) gives

$$0 = \sum_j \left(\mathrm{i}\hbar\,\dot{d}_j(t)\mathrm{e}^{-\mathrm{i}E_j t/\hbar}|j\rangle + d_j E_j \mathrm{e}^{-\mathrm{i}E_j t/\hbar}|j\rangle - d_j\mathrm{e}^{-\mathrm{i}E_j t/\hbar}\mathcal{H}^0|j\rangle - d_j\mathcal{H}^1(t)\mathrm{e}^{-\mathrm{i}E_j t/\hbar}|j\rangle\right) \tag{2.6}$$

or, after canceling the two middle terms,

$$\sum_j \mathrm{i}\hbar\,\dot{d}_j(t)\mathrm{e}^{-\mathrm{i}E_j t/\hbar}|j\rangle = \sum_j d_j\,\mathrm{e}^{-\mathrm{i}E_j t/\hbar}\mathcal{H}^1(t)|j\rangle. \tag{2.7}$$

We can pick out the time dependence of one of the coefficients, say d_k, by taking the inner product of $\langle k|$ with equation (2.7) giving

$$\mathrm{i}\hbar\,\dot{d}_k\mathrm{e}^{-\mathrm{i}E_k t/\hbar} = \sum_j d_j\mathrm{e}^{-\mathrm{i}E_j t/\hbar}\langle k|\mathcal{H}^1(t)|j\rangle, \tag{2.8}$$

which can be written as

$$\dot{d}_k = -\mathrm{i}\sum_j d_j\mathrm{e}^{\mathrm{i}\omega_{kj}t}\mathcal{H}^1_{kj}(t)/\hbar \tag{2.9}$$

where $\omega_{kj} = (E_k - E_j)/\hbar$ and $\mathcal{H}^1_{kj}(t) = \langle k|\mathcal{H}^1(t)|j\rangle$ are called the *matrix elements* of $\mathcal{H}^1$. Note that this equation is *exact*, and is really just the time-dependent Schrödinger equation in disguise.

2.2 Sudden jumps

As a first attempt at solving this equation, consider a really simple (indeed stupidly simple) model system, namely a two-level atom with a single electron which experiences an electric field $\boldsymbol{E}$ for a time τ. The perturbation Hamiltonian is then

$$\mathcal{H}^1 = -\boldsymbol{\mu}\cdot\boldsymbol{E} = \begin{cases} ezE & 0 \le t \le \tau \\ 0 & \text{otherwise} \end{cases} \tag{2.10}$$

where $\boldsymbol{\mu} = -e\boldsymbol{r}$ is the dipole moment of the atom arising from the separation of the electron and the nucleus, and the electric field direction has been taken as defining the z axis. From the symmetry of the atomic states it is obvious that

$$\langle g|\mathcal{H}^1|g\rangle = \langle e|\mathcal{H}^1|e\rangle = 0, \tag{2.11}$$

and we can choose to write

$$\langle g|\mathcal{H}^1|e\rangle = \hbar V \qquad \langle e|\mathcal{H}^1|g\rangle = \hbar V^* \tag{2.12}$$

where the second result is deduced from the first by the fact that the Hamiltonian is Hermitian, although in this case $V^* = V$.

Example 2.1 Calculate V for the transition between the $1s$ and $2p_z$ states in hydrogen. (The $2p_z$ state is the $2p$ state with $m_l = 0$.)

Solution

From above we have

$$V = \frac{eE}{\hbar}\langle \psi_{1,0,0}|z|\psi_{2,1,0}\rangle \tag{2.13}$$

where the low-lying hydrogen wavefunctions $\psi_{n,l,m}$ are given in Table 2.1. Writing $z = r\cos\theta$, the key integral is

$$\int_0^\infty dr \int_0^\pi d\theta \int_0^{2\pi} d\phi \sqrt{\frac{1}{\pi a_0^3}} e^{-r/a_0} \times r\cos\theta \times \sqrt{\frac{1}{32\pi a_0^5}} r\cos\theta\, e^{-r/2a_0} \times r^2\sin\theta. \tag{2.14}$$

Evaluating this integral gives

$$V = \frac{eE}{\hbar} \times \frac{128\sqrt{2}a_0}{243} \approx 0.74\frac{ea_0}{\hbar}E. \tag{2.15}$$

Table 2.1 The wavefunctions of the hydrogen atom take the form $\psi_{n,l,m}(r,\theta,\phi) = R_{n,l}(r)Y_{l,m}(\theta,\phi)$, where the low-lying radial functions $R_{n,l}$ and the spherical harmonics $Y_{l,m}$ are as given below

$R_{1,0}$	$\sqrt{\dfrac{4}{a_0^3}}e^{-r/a_0}$	$Y_{0,0}$	$\sqrt{\dfrac{1}{4\pi}}$
$R_{2,0}$	$\sqrt{\dfrac{1}{2a_0^3}}e^{-r/2a_0}\left(1-\dfrac{r}{2a_0}\right)$	$Y_{1,0}$	$\sqrt{\dfrac{6}{8\pi}}\cos\theta$
$R_{2,1}$	$\sqrt{\dfrac{1}{6a_0^3}}e^{-r/2a_0}\dfrac{r}{2a_0}$	$Y_{1,\pm1}$	$\mp\sqrt{\dfrac{3}{8\pi}}\sin\theta\, e^{\pm i\phi}$

The time dependence of the coefficients is given by

$$\dot{d}_e = -i\, d_g\, e^{i\omega_0 t}V, \tag{2.16}$$

$$\dot{d}_g = -i\, d_e\, e^{-i\omega_0 t}V, \tag{2.17}$$

where $\omega_0 = \omega_{eg} = -\omega_{ge}$ corresponds to the energy gap between the excited and ground states. These coupled differential equations can be solved by differentiating one equation with respect to time and substituting the other equation into the result, to give a single second-order ordinary differential equation. The procedure is fairly straightforward but messy. It is useful to start by considering the simplest case where the field is very strong, or the two energy levels are almost degenerate, so that $V \gg \omega_0$ and the exponential terms can simply be ignored. The equations

$$\dot{d}_e = -\mathrm{i}\, d_g\, V \qquad \dot{d}_g = -\mathrm{i}\, d_e\, V \qquad (2.18)$$

are now easy to solve; assuming the atom starts in the ground state (so that $d_g = 1$ and $d_e = 0$), the result is

$$d_g = \cos(V t) \qquad d_e = -\mathrm{i}\sin(V t). \qquad (2.19)$$

The effect of the sudden strong perturbation is to cause the system to make transitions from the ground state to the excited state and back again, with the amplitude of the excited state modulated sinusoidally at a rate given by V. The exact result has the same broad form: assuming that the atom starts in the ground state then

$$d_e = -\mathrm{i}\sqrt{\frac{4V^2}{4V^2 + \omega_0^2}}\ \sin\left(\frac{t\sqrt{4V^2 + \omega_0^2}}{2}\right)\mathrm{e}^{\mathrm{i}\omega_0 t/2}, \qquad (2.20)$$

which reduces to equation (2.19) when $\omega_0 \to 0$.

This sinusoidal modulation is called *Rabi flopping* and is also found in more realistic treatments of transitions. Note that flopping will only occur at all if the perturbation *connects* the two transitions, that is

$$V = \langle g|\mathcal{H}^1|e\rangle/\hbar \neq 0, \qquad (2.21)$$

and is only efficient if $V \gg \omega_0$, where $\hbar\omega_0$ corresponds to the gap between the energy levels. Thus a static field can be very effective at inducing transitions between degenerate energy levels, but will have little effect on non-degenerate levels unless it is very strong. In this latter case the field will cause transitions between many different pairs of levels, and the two-level atom assumption will not be valid. Indeed a sufficiently strong field will cause transitions to unbound states, effectively tearing the atom apart. Fortunately there are more subtle ways of inducing transitions.

2.3 Oscillating fields

A much more practical approach is to note that transitions can be induced by a small oscillating field, such as the electric field of a light wave, as long as the field is close to resonance with the desired transition. In many texts this result is derived using time-dependent perturbation theory, but it is more insightful to begin with an analytic result. Consider a co-sinusoidal oscillating electric field, with an angular frequency $\omega = 2\pi\nu$ and intensity $\mathcal{E}$; as the Schrödinger equation is linear this can be rewritten as the sum of two complex fields

$$\mathcal{E}(t) = \mathcal{E}\cos\omega t = \tfrac{1}{2}\mathcal{E}\left(\mathrm{e}^{\mathrm{i}\omega t} + \mathrm{e}^{-\mathrm{i}\omega t}\right) \qquad (2.22)$$

and for the moment we will only consider the first term in this sum and will ignore the *counter-rotating* component; justifications of this approach, which is called the *rotating wave approximation*, will be given below. The matrix elements of the perturbation Hamiltonian are now given by

$$\langle g|\mathcal{H}^1|e\rangle = \tfrac{1}{2}\hbar V \mathrm{e}^{\mathrm{i}\omega t}, \tag{2.23}$$

$$\langle e|\mathcal{H}^1|g\rangle = \left(\tfrac{1}{2}\hbar V \mathrm{e}^{\mathrm{i}\omega t}\right)^* = \tfrac{1}{2}\hbar V \mathrm{e}^{-\mathrm{i}\omega t}. \tag{2.24}$$

Inserting these into equation (2.9) gives for the time dependence of the coefficients

$$\dot{d}_e = -\tfrac{1}{2}\mathrm{i}\, d_g\, \mathrm{e}^{\mathrm{i}(\omega_0-\omega)t} V, \tag{2.25}$$

$$\dot{d}_g = -\tfrac{1}{2}\mathrm{i}\, d_e\, \mathrm{e}^{-\mathrm{i}(\omega_0-\omega)t} V, \tag{2.26}$$

which are exactly our previous results, except that ω_0 has now been replaced by $\omega_0 - \omega$, that is the difference between the frequency of the light and the resonance frequency of the system, and the strength of the perturbation has been halved. In particular, if the light is exactly resonant with the transition, so that $\omega_0 - \omega = 0$, then the simple results

$$d_g = \cos(Vt/2) \qquad d_e = -\mathrm{i}\sin(Vt/2) \tag{2.27}$$

are recovered. Thus Rabi flopping can be induced by a weak field oscillating in resonance with a transition. The populations of the ground and excited states are given by

$$P_g = \cos^2(Vt/2) = \tfrac{1}{2}[1 + \cos(Vt)], \tag{2.28}$$

$$P_e = \sin^2(Vt/2) = \tfrac{1}{2}[1 - \cos(Vt)] \tag{2.29}$$

and are sinusoidally modulated at a frequency V, called the *Rabi frequency*.

Note that the Rabi frequency refers to the rate of modulation of the *populations*, not the probability amplitudes, which are modulated at half this frequency. This can be seen as another example of *spinor* behavior: when an atom is coherently rotated from its ground state through an excited state and back to the ground state again its wavefunction picks up a sign of -1. There is, however, considerable variation among (and even within) textbooks as to whether V is taken as the strength of the *oscillating* field (as used here), or as the strength of the *rotating* field; this leads to minor variations in equations, and in particular in the formula for the Rabi frequency. Similarly some authors incorporate a factor of $\hbar$ into V rather than separating it out as done here.

This method can, of course, also be used to calculate the effects of off-resonance excitation, and the key results are implied above. However, more insight into this problem can be gained by using the rotating frame transformation and the vector model, which will be discussed in the next chapter. Although the discussion there is formally concerned with spins in magnetic fields, the results can, of course, be applied to any other two-level quantum system, as all two-level systems are fundamentally equivalent.

2.4 Time-dependent perturbation theory

Two assumptions were made in deriving the results above: firstly that we can treat the system as a two-level atom, and secondly that the counter-rotating component of the oscillating field can be ignored. This seems reasonable in light of the final result: as the counter-rotating component is far from resonance it will not be effective at inducing Rabi flopping unless the field is very strong. Furthermore, since the light will only induce transitions at frequencies close to ω it seems reasonable to ignore all other excited states. It might, however, be argued that this proof is circular, since the final result is assumed at the start.

To make a more rigorous argument it is necessary to return to equation (2.9), which is exact. We could solve this fairly easily for a two-level atom, but with an n-level system we will end up having to solve an nth-order differential equation. Furthermore, this equation will become extremely complex unless the form of the time-varying perturbation is extremely simple. To make further progress we will have to make approximations from the start, and if the perturbation is *small* then it makes sense to use a power series in $\mathcal{H}^1(t)$.

Consider a multi-level atom, and suppose that the system begins in some initial state $|i\rangle$ and we wish to obtain the amplitude of the system making a transition to some final state $|f\rangle$. The zero-order result is obtained by ignoring the perturbation completely (effectively setting $\mathcal{H}^1 = 0$), and substituting this into equation (2.9) gives the trivial result that the coefficients do not evolve. The first-order result is then obtained by using the zeroth-order wavefunction (that is, the unperturbed coefficients) with the first-order Hamiltonian, giving

$$\dot{d}_f = -\mathrm{i}\mathrm{e}^{\mathrm{i}\omega_{fi}t}\langle f|\mathcal{H}^1(t)|i\rangle/\hbar. \qquad (2.30)$$

(Note that at small times after the perturbation is first applied all the coefficients will be close to zero, except for d_i which will remain close to one, and we have assumed that $\langle i|\mathcal{H}^1|i\rangle = 0$ as before.) The solution is

$$d_f(t) = -\frac{\mathrm{i}}{\hbar}\int_0^t \mathrm{e}^{\mathrm{i}\omega_{fi}t'}\langle f|\mathcal{H}^1(t')|i\rangle\,\mathrm{d}t', \qquad (2.31)$$

where t' is just a dummy variable for the integration. This integral is, of course, a Fourier transform, suggesting that the process will be sensitive to components of $\mathcal{H}^1(t)$ oscillating near the frequency ω_{fi}. Furthermore, because the Fourier transform is *linear*, the total effect of applying several different perturbations is simply the sum of the effects of the individual perturbations. In particular it is possible to treat an oscillating perturbation as the sum of two counter-rotating perturbations, equation (2.22), and it is possible to treat *any* perturbation as a sum of oscillating terms. For a single oscillating term with angular frequency ω the solution is

$$d_f(t) = -\frac{i}{\hbar} \int_0^t e^{i(\omega_{fi}-\omega)t'} \hbar V(\omega)\, dt' \tag{2.32}$$

$$= -iV(\omega) \times \left[\frac{e^{i(\omega_{fi}-\omega)t'}}{i(\omega_{fi}-\omega)}\right]_0^t \tag{2.33}$$

$$= -iV(\omega)e^{i(\omega_{fi}-\omega)t/2} \times \left(\frac{e^{i(\omega_{fi}-\omega)t/2} - e^{-i(\omega_{fi}-\omega)t/2}}{i(\omega_{fi}-\omega)}\right) \tag{2.34}$$

$$= -iV(\omega)e^{i(\omega_{fi}-\omega)t/2} \times \left(\frac{2i\sin[(\omega_{fi}-\omega)t/2]}{2i(\omega_{fi}-\omega)/2}\right) \tag{2.35}$$

$$= -iV(\omega)e^{i(\omega_{fi}-\omega)t/2} \operatorname{sinc}[(\omega_{fi}-\omega)t/2] \times t, \tag{2.36}$$

where $\operatorname{sinc}(x) = \sin(x)/x$. The sinc function arises naturally whenever a Fourier transform is taken of an oscillation with a finite extent, and can be considered as measuring the uncertainty in the frequency of the oscillation.

First, consider the case when the oscillation is exactly resonant with the transition, so that $\omega_{fi} - \omega = 0$. Since $\operatorname{sinc}(0) = 1$, equation (2.36) reduces to

$$d_f(t) = -iVt \tag{2.37}$$

and this result is identical to the previous result for a two-level atom, equation (2.19), at short times when $\sin(Vt) \approx Vt$. At longer times the treatment breaks down, as it is no longer reasonable to assume that d_i is always one. This can be overcome by using higher orders of perturbation theory: the conceptually simplest method is to feed the first-order wavefunctions back into the algorithm to obtain second-order wavefunctions, and so on, ultimately giving series expansions of the underlying sine and cosine modulations.

Equation (2.36) can also be used to look at the effects of excitation away from resonance. This will be identical to excitation on-resonance, except that the strength of the interaction is scaled down by $\operatorname{sinc}[(\omega_{fi}-\omega)t/2]$. Clearly the effect will be very small unless ω is close to resonance, justifying our previous decision to use the two-level atom model and to ignore the counter-rotating component of the excitation field. More interestingly, this result shows that excitation becomes more "choosy" as time goes on; this is not particularly surprising, however, as it simply reflects the fact that the frequency of an oscillation becomes better defined as it is observed or applied over a long period.

2.5 Rabi flopping and Fermi's Golden Rule

The treatment above looks good, but clashes with common experience. Consider the effect of on-resonance excitation, equation (2.37), and calculate how the population of the final state varies with time. Since $P_f = |d_f|^2$ this is given by

$$P_f(t) = V^2 t^2, \tag{2.38}$$

so the degree of excitation varies *quadratically* with time, or, equivalently, the rate at which transitions occur increases linearly with time. In fact, however, the excited state population is often observed to grow *linearly* with time, so that the transition rate is constant.

This apparent discrepancy is easily explained. So far we have assumed that the energy levels of an atom are perfectly sharp, so that any transition has a single exact frequency, but this is quite untrue. Every excited state of an atom has a finite lifetime (ultimately limited by the spontaneous emission lifetime), and so has a corresponding uncertainty in its energy; thus the frequency of a transition is not in fact well defined! It is, therefore, usually necessary to integrate the transition probability over the whole range of transition frequencies, and when this is done it is found that

$$P_f(t) \propto V^2 t \tag{2.39}$$

in agreement with naive expectations. A detailed derivation of this result, known as Fermi's Golden Rule, can be found in many standard texts.

Which then is right? Does light cause an atom to undergo Rabi flopping, or does excitation follow Fermi's Golden Rule? This question can be considered from the point of view of theory and from that of experiment.

The essential reason underlying the linear behavior in equation (2.39) is easy to understand. As previously noted, the system becomes increasingly choosy about whether or not to make a transition as time goes on, and the increasing fussiness counteracts the intrinsic tendency of the transition rate to grow, resulting in a constant transition rate overall. This effect is only important, however, for times which are long in comparison with the inverse of the width of the transition; in effect this means times which are long in comparison with the lifetime of the excited state. The time for which the light is applied will obviously depend on the time it takes to have a significant effect, which is conveniently parameterized by the oscillation frequency in equation (2.19). We can thus distinguish two extreme regimes of behavior, depending on V and the state lifetime τ:

1. Strongly driven transitions: $V\tau \gg 1$. In this case the system undergoes Rabi oscillations between the ground and excited state. This case is sometimes called *coherent control*, and is suitable for quantum information processing experiments.
2. Weakly driven transitions: $V\tau \ll 1$. In this case the system obeys Fermi's Golden Rule and a constant transition probability is observed. The long-term behavior of the system is described by *rate equations*, which explicitly include the effects of spontaneous emission from the excited state.

For transitions at optical frequencies, the lifetimes of the excited state are usually fairly short, and transitions are usually weakly driven, although it is possible to observe Rabi oscillation behavior, either by using very high laser powers (for example, by focusing a laser down to a very small spot), or by artificially suppressing spontaneous decay (this can be achieved by placing the atom in a high-finesse optical cavity, so that the system can only emit photons into the resonant modes of the cavity, and ensuring that none of these modes are resonant with the atomic transition). Thus direct optical transitions are not normally suitable candidates for coherent quantum control. The obvious solution is to use transitions to a low-lying quantum state, so that the transitions occur at much lower frequencies. (From the Einstein A and B coefficients we know that the relative importance of spontaneous decay and driven transitions goes as the third power of the frequency.) There are, however, two problems with this approach.

Firstly, most transitions from the ground state to low-lying excited states are *forbidden*, that is the matrix element for the transition $\mathcal{H}^1_{eg} = 0$. Although working out the exact matrix element connecting two states for a given perturbation can be quite complicated, it is relatively simple to list *selection rules* which determine whether it is zero (a forbidden transition) or non-zero (an allowed transition). So far we have been considering transitions induced by the interaction between the electric field of light and the electric dipole moment of an atom, and so it is more accurate to state that most transitions from the ground state to low-lying excited states are *electric dipole forbidden*. It is, of course, possible to find many low-frequency transitions between pairs of excited states which are electric dipole allowed, but in this case *both* states will be broadened by spontaneous emission, and the possibility of coherent control is further suppressed.

Example 2.2   We have already worked out the matrix element $\langle \psi_{100}|z|\psi_{210}\rangle$. We can show that $\langle \psi_{100}|z|\psi_{211}\rangle = 0$ by noting that the integral over ϕ is obviously zero, and so the whole integral must also be zero. Similarly $\langle \psi_{100}|z|\psi_{200}\rangle = 0$ because of the integral over θ.

A second problem is that low-frequency light has a long wavelength, and this makes it difficult to focus. For quantum information processing it is usually necessary to excite one atom without exciting another similar atom which is physically close by. Such selective excitation can only be achieved if the light can be focused down to a spot which is small compared with the separation of the atoms, and the minimum spot size, given by the Abbe limit

$$d \gtrsim \lambda/2, \tag{2.40}$$

is limited by the wavelength of the light. For visible light this resolution limit will be around 300 nm, but for 1 GHz radiation the limiting separation will be around 15 cm.

2.6   Raman transitions

The solution to both these problems is to use *Raman* transitions to connect two low-lying energy levels. Many textbooks do not discuss Raman transitions at all, and many of the rest only discuss the *Raman effect* used in spectroscopy, rather than coherent Raman transitions. Fortunately the basic idea is easy to understand.

Consider a system with three energy levels, $|g\rangle$ and $|e\rangle$, which form the basis of our qubit, and an additional level $|a\rangle$. We will assume that transitions between $|g\rangle$ and $|e\rangle$ are forbidden, but that both of these can make transitions to $|a\rangle$. Suppose the system is illuminated by two lasers, one in resonance with each of the two allowed transitions. The result of this process will be a complicated evolution of the system between the three states, with transfers from

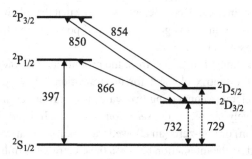

Fig. 2.1 Some energy levels and transitions in ^{40}Ca$^+$ ions. Allowed transitions are shown as solid lines, while forbidden transitions are shown as dashed lines, with the wavelength given in nanometers (nm).

$|g\rangle$ to $|e\rangle$ occurring via the additional state $|a\rangle$. This (in principle) solves the problem of making forbidden transitions, but is not an effective solution for two reasons. Firstly, the system cannot go from $|g\rangle$ to $|e\rangle$ without passing through $|a\rangle$, and thus we no longer have a proper two-level system, and secondly, it remains hard to drive the system strongly enough that Rabi behavior occurs.

The solution is to tune both lasers *away* from the frequencies of the two allowed transitions by the same amount, so that the energy difference between photons in the two beams still matches the energy gap between $|g\rangle$ and $|e\rangle$. The remarkable result is that although the two allowed transitions no longer occur, Rabi flopping occurs for the *forbidden* transition between $|g\rangle$ and $|e\rangle$. This is an example of a *two-photon* process: in effect a photon is absorbed from one laser beam, while the other beam stimulates the emission of a second photon. The transition is sometimes described as occurring via a *virtual state*, but in fact occurs via off-resonance interactions with the (real) state $|a\rangle$. Because these transitions are off-resonance, the Rabi frequency is scaled down from its naive value V by a factor V/Δ, where Δ is the frequency offset from resonance,[1] but this is not a major problem as the laser power can be increased by a corresponding factor. An important advantage is that the system can be driven strongly, as the relevant state lifetimes τ (see the previous section) are those of $|g\rangle$ and $|e\rangle$; the lifetime of the additional state $|a\rangle$ is irrelevant as this state is never populated! Raman transitions provide an almost ideal solution to the problem of inducing Rabi flopping between atomic energy levels, and are very commonly used.

Example 2.3 Figure 2.1 shows schematically the energy-level structure in ^{40}Ca$^+$ ions, which provides a simple system for demonstrating the basic ideas in trapped ion implementations of quantum computing (^{40}Ca has nuclear spin $I = 0$, and so there is no hyperfine structure to worry about). A qubit could be implemented using the ^{2}D$_{3/2}$ and ^{2}D$_{5/2}$ levels as $|0\rangle$ and $|1\rangle$; these are relatively long lived as their transitions down to the ^{2}S$_{1/2}$ ground state are forbidden. The direct transition between $|0\rangle$ and $|1\rangle$ is forbidden, but can be driven using a

[1] This assumes that the two underlying transitions have the same matrix element, V. In general these will differ, and the Rabi frequency depends on $V_1 V_2/\Delta$; the exact form depends on the choice of conventions as usual.

Raman transition with $^2P_{3/2}$ as the third level. The other transitions shown can be used for initialization and readout, as explained later on.

Example 2.4 In reality things are significantly more complicated than this, as all these levels are split into several components, corresponding to different values of the magnetic quantum number m_j. It is necessary to lift the degeneracy of these levels, using a magnetic field which splits them due to the Zeeman effect, and then use single components and single transitions to implement qubits and gates. Experimental implementations based on $^{40}Ca^+$ ions either use the two $m_j = \pm\frac{1}{2}$ components of the $^2S_{1/2}$ level, or use two components in the $^2S_{1/2}$ and $^2D_{5/2}$ levels; the "forbidden" transition between these levels at 729 nm is weakly allowed by quadrupole effects and can be excited with intense fields.

2.7 Rabi flopping and Ramsey fringes

The underlying Hamiltonian describing Rabi flopping, whether achieved directly with resonant radiation or indirectly with Raman transitions, can be written as

$$\mathcal{H} = \hbar \times \begin{pmatrix} 0 & V/2 \\ V/2 & 0 \end{pmatrix} = \hbar V \, \sigma_x/2, \tag{2.41}$$

where V is the Rabi frequency. Neglecting global phases, the corresponding propagator is

$$U = \begin{pmatrix} \cos(Vt/2) & -i\sin(Vt/2) \\ -i\sin(Vt/2) & \cos(Vt/2) \end{pmatrix}. \tag{2.42}$$

This propagator has already been examined in some detail, but it can be viewed in a quite different way, namely as the nth-POWER-OF-NOT quantum logic gate, with $n = Vt/\pi$. When $Vt = \pi$ (a π-pulse, or 180° pulse, with $n = 1$) we get a NOT gate, which interconverts $|0\rangle$ and $|1\rangle$, while a 90° pulse ($n = \frac{1}{2}$) produces equally weighted superpositions of $|0\rangle$ and $|1\rangle$.

In order to implement general single-qubit gates it is also useful to be able to implement Hamiltonians proportional to σ_y. This can be achieved by simply altering the *phase* of the radiation, so that the perturbation takes the form $\hbar V \cos(\omega t + \phi)$. The effective Hamiltonian can then be written as

$$\mathcal{H} = \hbar \times \begin{pmatrix} 0 & e^{-i\phi}V/2 \\ e^{i\phi}V/2 & 0 \end{pmatrix} = \hbar V \, (\sigma_x \cos\phi + \sigma_y \sin\phi)/2, \tag{2.43}$$

where we have assumed the radiation is applied on-resonance and made the rotating wave approximation as usual. Thus by appropriate choice of ϕ we can generate Hamiltonians proportional to σ_x, or σ_y, or at any angle between them. If we take the case $\phi = \pi/2$, so that $\mathcal{H} \propto \sigma_y$, then the evolution propagator is

$$U = \begin{pmatrix} \cos(Vt/2) & -\sin(Vt/2) \\ \sin(Vt/2) & \cos(Vt/2) \end{pmatrix}. \tag{2.44}$$

Clearly the system undergoes Rabi oscillations at the same frequency as before.

If only populations are considered then the phase of the radiation has no effect, but if the amplitudes of the two states are considered then the phase is important. Of course the *absolute phase* of an oscillation is essentially meaningless, with only the relative phase of two oscillations being well defined. Thus it is not possible to define an absolute phase for a single Rabi pulse, but it is possible to define a relative phase for two or more pulses. This will be briefly explored below.

Ramsey fringes occur when two 90° pulses are applied to a two-level quantum system, separated by a time period during which the system is allowed to undergo free evolution. The overall result is an oscillation with a frequency depending on the energy gap between the two levels of the system, and this is most easily analyzed using a gate network. The details depend on the exact form of the 90° pulses, and in particular what phase they have, but the simplest situation occurs when they can be treated as Hadamard gates; as shown in Section 1.4, these are closely related to 90° pulses. Next we must consider how to represent the period of free evolution, during which the system evolves under the background Hamiltonian, that is the Hamiltonian in the absence of any applied fields. We have assumed that our basis states are eigenstates, and so the Hamiltonian must be diagonal in this basis. Placing the energy zero mid-way between the energies of the two states $|0\rangle$ and $|1\rangle$ it takes the form

$$\mathcal{H} = \begin{pmatrix} -\frac{1}{2}\Delta E & 0 \\ 0 & \frac{1}{2}\Delta E \end{pmatrix} = -\hbar\omega_0\,\sigma_z/2, \qquad (2.45)$$

where $\Delta E = E_1 - E_0$ and $\omega_0 = \Delta E/\hbar$ as usual. The propagator describing the free evolution is

$$U = \exp(-\mathrm{i}\mathcal{H}t/\hbar) = \exp(\mathrm{i}\omega_0 t\,\sigma_z/2) = \cos(\omega_0 t/2)\sigma_0 + \mathrm{i}\sin(\omega_0 t/2)\sigma_z, \qquad (2.46)$$

where the last step uses equation (1.19).

The experiment can now be described by a gate network, and using the linearity of matrix operations

$$\mathrm{H}U\mathrm{H} = \cos(\omega_0 t/2)\mathrm{H}\sigma_0\mathrm{H} + \mathrm{i}\sin(\omega_0 t/2)\mathrm{H}\sigma_z\mathrm{H}. \qquad (2.47)$$

This expression can be simplified using $\mathrm{H}\sigma_0\mathrm{H} = \mathrm{H}^2 = \sigma_0$ and $\mathrm{H}\sigma_z\mathrm{H} = \sigma_x$ to obtain

$$\mathrm{H}U\mathrm{H} = \cos(\omega_0 t/2)\sigma_0 + \mathrm{i}\sin(\omega_0 t/2)\sigma_x = \exp(\mathrm{i}\omega_0 t\,\sigma_x/2), \qquad (2.48)$$

showing that the overall effect of the sequence is to perform Rabi flopping at a frequency which depends on the *internal frequency*, ω_0, of the system.

So why is this experiment known as Ramsey fringes? Suppose that the system starts off in the ground state $|0\rangle$, and after a time t we measure the state in the computational basis. The probability that the system is still in the ground state is then $\cos^2(\omega_0 t/2)$. In other words the signal is *time-modulated*, and this time modulation takes exactly the same form as the spatial modulation fringes seen in a two-slit experiment. The more general case, when the two pulses are arbitrary 90° pulses, is very similar. The results are essentially unchanged, but the phase of the fringes is shifted, with the phase angle depending on the relative phase of the two pulses; the case of two Hadamard gates is recovered when the phases of the two pulses are 180° apart.

2.8 Measurement and initialization

Measuring the quantum state of an atom or ion is in principle relatively straightforward, as it is possible to make transitions to states separated in energy by quite large energies (several electronvolts), and these transitions can be detected, for example by detecting photons emitted by fluorescence. The process can become quite complex, and is explored in more detail in Section 10.9 in Part II. Preparing atoms in desired initial states is also straightforward in principle as the energy gaps involved may be large enough to allow direct cooling to the ground state at reasonable temperatures. In many cases, however, the qubit is implemented using two hyperfine levels of the ground state, and in this case it is necessary to use more sophisticated approaches based on optical pumping, as described in Section 10.3.

Further reading

The theory of Rabi flopping and time-dependent perturbation theory, including Fermi's Golden Rule, is well covered in standard texts on quantum mechanics, such as Gasiorowicz (2003) and Binney and Skinner (2010), or atomic physics, such as Budker *et al.* (2004) and Foot (2005). The definitive text (Cohen-Tannoudji *et al.*, 1992) is extremely challenging at this level. The Abbe limit, and techniques by which it can in principle be overcome, are discussed in optics texts such as Hecht (2002) and Lipson *et al.* (2011). Quantum optics is well described in many texts, including Vedral (2005) and Gerry and Knight (2005).

A basic introduction to ion traps can be found in Ozeri (2011); more detailed sources will be listed in Part II.

Exercises

2.1   Explain why the selection rules derived for hydrogen atoms can also be applied to ions with a single electron in their outer shell, such as Ca^+.

2.2   Consider the possibility of using the $^2S_{1/2}$ and $^2P_{1/2}$ to encode a qubit in $^{40}Ca^+$ ions. For simplicity we assume that the matrix element $\langle z \rangle = \langle \psi_i | z | \psi_f \rangle \sim a_0$ for allowed transitions, and is zero for forbidden transitions. Calculate the spontaneous decay time of this transition using $1/\Gamma = (3\pi \epsilon_0 \hbar c^3)/(\omega^3 e^2 \langle z \rangle^2)$, and estimate the electric field strength needed to perform Rabi flopping on this transition using a resonant oscillating electric field.

2.3   Suppose we tried to excite this transition by brute force, using a very large jump in a static electric field. Estimate the field strength required to make this work, and comment on your result.

2.4   Estimate the limiting spatial resolution in this system (you may assume the Abbe limit).

2.5   Comment on the expected excited state population at 300 K.

2.6   The peak electric field in a laser beam can be calculated using $E_p = 2\sqrt{Pc\mu_0/A}$, where P is the power of the laser and A is the cross-sectional area of the beam. Estimate the laser power required to perform Rabi flopping assuming the laser beam is focused to a uniform spot with a diameter given by the Abbe limit.

Spins in magnetic fields

Spins in magnetic fields provide one of the simplest and most natural physical systems for implementing quantum bits; indeed the relationship between a spin and a qubit is so close that the terms are sometimes used interchangeably. Experimental spin physics is not often studied in physics courses, which is a pity, as it provides one of the simplest examples of coherent quantum control available. The treatment is essentially identical to that of two-level atoms in laser fields, except that transitions can almost always be treated as strongly driven. The language traditionally used to describe spin systems is, however, quite different from that used to describe atomic transitions.

For simplicity the discussion below will largely assume that the spins in question are nuclear spins rather than electron spins. The underlying physics of the two systems is very similar, and the same approach can in principle be used in both cases, but in practice systems of electron spins frequently suffer from additional complicating factors which we do not consider here.

3.1 The nuclear spin Hamiltonian

Just like electrons, atomic nuclei possess an intrinsic angular momentum, called spin. This arises from the coupling between the intrinsic spins of the protons and neutrons making up the nucleus. A nucleus with spin quantum number I has spin angular momentum $\hbar I$ and an associated magnetic moment $\boldsymbol{\mu}$, given by

$$\boldsymbol{\mu} = \gamma \hbar \boldsymbol{I}, \tag{3.1}$$

where γ is called the *gyromagnetic ratio* of the nucleus. Although it is in theory possible to calculate these properties from first principles, it is usually best simply to treat the details of nuclear spins as experimentally measured quantities.

If the spin is placed in a magnetic field $\boldsymbol{B}$ then the interaction between the magnetic moment and the field is described by the Zeeman Hamiltonian

$$\mathcal{H} = -\boldsymbol{\mu} \cdot \boldsymbol{B} \tag{3.2}$$

and the standard convention is to orient the z axis along the magnetic field, so that

$$\mathcal{H} = -\mu_z B = -\hbar \gamma B I_z = -\hbar \omega_L I_z, \tag{3.3}$$

where I_z is the projection of $\boldsymbol{I}$ onto the z axis and ω_L is called the *Larmor frequency*. As this is a quantum mechanical system, I_z cannot take on any value, but only those values between $-I$ and I in integer steps. The simplest situation occurs for a spin-$\frac{1}{2}$ nucleus, in which case there are only two possible values, $I_z = \pm\frac{1}{2}$. The most important example of a spin-$\frac{1}{2}$ nucleus is the hydrogen (^{1}H) nucleus, but many others exist, most notably ^{13}C, ^{15}N, ^{19}F and ^{31}P.

The effect of a magnetic field on a spin-$\frac{1}{2}$ nucleus is to split apart the two spin states, with a splitting $\hbar\omega_L$, and these two energy levels provide an obvious implementation of a qubit. There is enormous variation in the notation used to describe these two spin states in the literature. Some relatively common notations are to call the spin states α and β, or to call them spin-up ($\uparrow$) and spin-down ($\downarrow$), or simply to call them $|\frac{1}{2}\rangle$ and $|-\frac{1}{2}\rangle$. As usual we will avoid these arguments by calling the two states $|0\rangle$ and $|1\rangle$ or $|g\rangle$ and $|e\rangle$. Note that in this case the system really does have only two levels, and so we do not need to make a two-level approximation.[1]

The transitions between these two spin states are *electric dipole forbidden*, as they violate the electric dipole selection rule $\Delta M_S = 0$, but they can be induced by magnetic fields. Another way of looking at this is that the electric field matrix element $\langle 1|\mathbf{E}|0\rangle = 0$, while the magnetic field matrix element $\langle 1|\mathbf{B}|0\rangle$ will be non-zero as long as the magnetic field is not parallel to the z axis. Thus if a strong magnetic field is suddenly applied at right angles to the main magnetic field then transitions between the two spin states will occur. More realistically, the same effect can be achieved by applying a weak oscillating magnetic field as long as it oscillates in resonance with the transition, that is at the Larmor frequency.

Spins in magnetic fields are, in some sense, a more natural quantum mechanical system than atoms in a laser field. The reason for this is not any fundamental property of the two systems, but simply a matter of practicalities: the most important difference is simply that the transitions between nuclear spin energy levels occur at very much lower frequencies. It is also possible to use electron spins as qubits, and the basic techniques are very similar. Electron transitions usually occur at much higher frequencies than nuclear transitions, as electrons have a much larger magnetic moment, but the frequencies are still very low compared with direct atomic transitions.

Nuclear spin transition frequencies depend both on the magnetic field strength used and on intrinsic properties of the nuclei. The largest static magnetic fields currently available[2] are around 20 T, and the most sensitive of the stable nuclei is ^{1}H (hydrogen), for which transition frequencies in the range up to 1 GHz are found, corresponding to the radio-frequency (RF) portion of the spectrum. Most studies of ^{1}H take place at frequencies in the range 400–800 MHz, and studies of other nuclei (except for ^{19}F, ^{3}He and the dangerously radioactive nucleus ^{3}H) take place at significantly lower frequencies.

---

[1] There is a subtlety here which sometimes traps the unwary. The angular momentum of the two eigenstates of a spin-$\frac{1}{2}$ particle, $|\frac{1}{2}\rangle$ and $|-\frac{1}{2}\rangle$, is not completely aligned with the corresponding quantization axis, as the projection of the angular momentum onto this axis take the values $\pm 1/2\,\hbar$, while the magnitude of the angular momentum is $\sqrt{3}/2\,\hbar$. For this reason spin states are sometimes drawn as cones centered on the quantization axis. However the corresponding Bloch vectors do lie exactly along the $\pm z$ axes of the Bloch sphere, and the cone picture is best forgotten when thinking about quantum information.

[2] These fields are achieved using superconducting electromagnets, and are limited by the *critical field* and *critical current* of the superconducting wires; larger fields are available for *short* periods of time by using pulsed electromagnets, and extremely large fields are available for very short times using destructive techniques.

There are two principal advantages of working with RF. The first is that spontaneous emission rates at these low frequencies are completely negligible, and so it should be easy to reach the coherent control region. The second advantage is that RF radiation is extremely easy to generate and control. While experimentalists working with lasers have to work hard to control the frequency, amplitude and phase of laser light, any desired RF pattern can be obtained simply by asking a computer to generate it. For this reason coherent control of nuclear spins has flourished for decades under the name of *nuclear magnetic resonance* or NMR.

There are, however, two major disadvantages of working with RF. The first is that the wavelength of RF radiation is so large that spatially selective excitation is essentially impossible, as discussed previously. The second is that the energy of RF photons is so small that it is virtually impossible to detect single photons. Both of these problems have major consequences for the use of NMR as an implementation of quantum information processing.

3.2 The rotating frame

Transitions between nuclear spin states can be treated using exactly the same techniques as we used previously to study transitions in a two-level atom, but it is more common to use a subtly different (though ultimately equivalent) approach, based on transforming the problem into a *rotating frame*. Consider a general wavefunction $|\psi\rangle$, which we choose to write as

$$|\psi\rangle = U|\tilde{\psi}\rangle, \tag{3.4}$$

where U simply describes the transformation between two different bases which can be used to describe the wavefunction. Note that

$$|\tilde{\psi}\rangle = U^{-1}|\psi\rangle = U^{\dagger}|\psi\rangle, \tag{3.5}$$

where we have used the fact that basis-state transformations are unitary. If we transform the wavefunction into a new basis then we must also transform the Hamiltonian, and this transformation can be worked out using the time-dependent Schrödinger equation

$$i\hbar\frac{\partial}{\partial t}|\psi\rangle = \mathcal{H}|\psi\rangle. \tag{3.6}$$

In the transformed basis

$$i\hbar\frac{\partial}{\partial t}|\tilde{\psi}\rangle = i\hbar\frac{\partial}{\partial t}\left(U^{\dagger}|\psi\rangle\right) \tag{3.7}$$

$$= i\hbar\left[U^{\dagger}\frac{\partial}{\partial t}|\psi\rangle + \left(\frac{\partial U^{\dagger}}{\partial t}\right)|\psi\rangle\right] \tag{3.8}$$

$$= \left[U^{\dagger}\mathcal{H} + i\hbar\left(\frac{\partial U^{\dagger}}{\partial t}\right)\right]|\psi\rangle. \tag{3.9}$$

Using equations (3.4) and (3.6) gives

$$i\hbar\frac{\partial}{\partial t}|\tilde{\psi}\rangle = \left[U^\dagger\mathcal{H}U + i\hbar\left(\frac{\partial U^\dagger}{\partial t}\right)U\right]|\tilde{\psi}\rangle = \tilde{\mathcal{H}}|\tilde{\psi}\rangle, \tag{3.10}$$

and so the transformed Hamiltonian has the form

$$\tilde{\mathcal{H}} = \left[U^\dagger\mathcal{H}U + i\hbar\left(\frac{\partial U^\dagger}{\partial t}\right)U\right]. \tag{3.11}$$

The first term in the transformed Hamiltonian is simply the obvious transformation of $\mathcal{H}$ into the new basis, but the second term is more subtle. This term is zero for fixed transformations, and corresponds to a *fictitious energy*, which is analogous to the fictitious forces that arise in classical mechanics when working in accelerating frames.

To take a concrete example, consider a spin-$\frac{1}{2}$ particle in a static magnetic field along the z axis and experiencing an oscillating magnetic field at right angles. This oscillating field can be achieved by using the magnetic component of an appropriate electromagnetic field oscillating at the resonance frequency, which corresponds to RF. Thus the Hamiltonian can be written in matrix form as

$$\mathcal{H} = \begin{pmatrix} -\frac{1}{2}\hbar\omega_0 & \hbar V\cos\omega t \\ \hbar V\cos\omega t & \frac{1}{2}\hbar\omega_0 \end{pmatrix}, \tag{3.12}$$

where we have used $|g\rangle$ and $|e\rangle$ as our basis states and have chosen to place the energy zero half-way between our two states. We then choose the transformation

$$U = \begin{pmatrix} e^{i\omega t/2} & 0 \\ 0 & e^{-i\omega t/2} \end{pmatrix}, \tag{3.13}$$

which corresponds to using basis states that rotate in synchrony with one component of the oscillating field. Applying equation (3.11), the Hamiltonian in this new frame is

$$\tilde{\mathcal{H}} = \begin{pmatrix} \frac{1}{2}\hbar(\omega-\omega_0) & \hbar V\cos(\omega t)e^{-i\omega t} \\ \hbar V\cos(\omega t)e^{i\omega t} & -\frac{1}{2}\hbar(\omega-\omega_0) \end{pmatrix}. \tag{3.14}$$

Next we define the *detuning* as $\delta = \omega - \omega_0$ and separate the oscillating term into two counter-rotating terms. Finally we apply the rotating wave approximation as before, and simply ignore the rapidly varying terms. Thus to a good approximation[3]

$$\tilde{\mathcal{H}} = \begin{pmatrix} \frac{1}{2}\hbar\delta & \frac{1}{2}\hbar V \\ \frac{1}{2}\hbar V & -\frac{1}{2}\hbar\delta \end{pmatrix}. \tag{3.15}$$

It is important to remember that although this result has been derived for the case of a spin in a magnetic field, the method is entirely general, and an identical result could have been derived for an atom in a laser field: all two-level quantum systems (qubits) are basically the same.

---

[3] The rotating wave approximation is not quite so good an approximation in this case as it was for transitions between atomic energy levels as the frequencies involved are much lower. Careful calculations indicate that the counter-rotating component gives rise to a small shift in the transition frequencies known as a Bloch–Siegert shift, which is related to the AC Stark shift.

3.3 On- and off-resonance excitation

As usual the simplest case occurs when the excitation is on-resonance, so that $\delta = 0$ and the Hamiltonian in the rotating frame is

$$\tilde{\mathcal{H}} = \begin{pmatrix} 0 & \frac{1}{2}\hbar V \\ \frac{1}{2}\hbar V & 0 \end{pmatrix} \tag{3.16}$$

which is clearly related to one of the Pauli matrices, that is

$$\tilde{\mathcal{H}} = \tfrac{1}{2}\hbar V \sigma_x. \tag{3.17}$$

We can calculate the propagator describing the evolution under this Hamiltonian to get

$$\tilde{U} = \exp(-\mathrm{i}\tilde{\mathcal{H}}t/\hbar) = \exp(-\mathrm{i}\theta\sigma_x/2), \tag{3.18}$$

where $\theta = Vt$. Now we have previously derived a formula for the matrix exponential of σ_x, equation (1.19), and so we know that

$$\tilde{U} = \begin{pmatrix} \cos(Vt/2) & -\mathrm{i}\sin(Vt/2) \\ -\mathrm{i}\sin(Vt/2) & \cos(Vt/2) \end{pmatrix}. \tag{3.19}$$

If the system starts off in the ground state $|g\rangle$ then the state at later times is given by

$$\tilde{\psi} = \tilde{U}\begin{pmatrix} 1 \\ 0 \end{pmatrix} = \begin{pmatrix} \cos(Vt/2) \\ -\mathrm{i}\sin(Vt/2) \end{pmatrix}, \tag{3.20}$$

in complete agreement with equation (2.27). Choosing $Vt = \pi$ will interconvert the ground and excited states, that is implement a NOT gate.

None of this should be surprising: it is all exactly as expected from the discussion in Sections 1.3 and 2.7 where we considered how to implement quantum logic gates, and showed that a NOT gate could be implemented by applying a Hamiltonian proportional to σ_x for an appropriate time. As described in the previous chapter, more general quantum logic gates can be implemented by varying the time t and by controlling the phase of the excitation.

Next we consider the case when the radiation is not quite in resonance with the transition frequency, so that the Hamiltonian takes the general form, equation (3.15). The propagator is then

$$\tilde{U} = \exp\left[-\mathrm{i} \times \begin{pmatrix} \delta/2 & V/2 \\ V/2 & -\delta/2 \end{pmatrix} \times t\right] \tag{3.21}$$

and brute force calculation (using a symbolic mathematics program is a great help with calculations of this kind) gives the result

$$\tilde{U} = \begin{pmatrix} \cos(\Omega t/2) - \mathrm{i}(\delta/\Omega)\sin(\Omega t/2) & -\mathrm{i}(V/\Omega)\sin(\Omega t/2) \\ -\mathrm{i}(V/\Omega)\sin(\Omega t/2) & \cos(\Omega t/2) + \mathrm{i}(\delta/\Omega)\sin(\Omega t/2) \end{pmatrix}, \tag{3.22}$$

where $\Omega = \sqrt{V^2 + \delta^2}$. Note that on-resonance, $\delta = 0$ and $\Omega = V$, and our previous results are recovered. Off-resonance, we see that the frequency of the Rabi oscillations is increased, so that $\Omega > V$, but the efficiency is reduced, so that $|0\rangle$ cannot be completely converted to $|1\rangle$.

Seen from the conventional point of view, off-resonance excitation is a bad thing, but from the viewpoint of quantum control it can provide a direct route to certain quantum logic gates. An important example occurs in the case when $V = \delta$, so that $\Omega = \sqrt{2}V$. Choosing t such that $\Omega t/2 = \pi/2$ gives

$$\tilde{U} = -\mathrm{i} \times \begin{pmatrix} \frac{1}{\sqrt{2}} & \frac{1}{\sqrt{2}} \\ \frac{1}{\sqrt{2}} & \frac{-1}{\sqrt{2}} \end{pmatrix} \tag{3.23}$$

which (neglecting an irrelevant global phase) is the Hadamard gate, one of the most important single-qubit logic gates. However, it is usually simpler in practice to consider only on-resonance excitation, and to construct gates such as the Hadamard gate using gate networks as described in Section 1.4.

Note that in the treatment used here the background Hamiltonian (that is, the Hamiltonian when $V = 0$) appears to change sign upon moving into the rotating frame [compare equations (3.12) and (3.15)]. Of course this is simply a consequence of the notation used, and has no fundamental significance. It is, however, necessary to be careful in reading treatments of NMR as the signs of frequencies are frequently treated in a casual and inconsistent fashion.

3.4 The vector model

There is another way of looking at spins in magnetic fields, usually called the vector model, which was developed by Bloch and is very widely used in the field of NMR. This is an entirely classical method for thinking about the situation, but it does give an accurate description of a single isolated spin. By the obvious extension it can also be used to describe any other single-qubit system.

We have already seen that the state of a spin can be represented as a Bloch vector, pointing from the origin to an appropriate point on the Bloch sphere. The vector model represents a spin by a classical magnetic moment pointing along this vector. If the spin is placed in a magnetic field along the z axis then it will *precess* around the field, at the Larmor frequency, which depends on the strength of the magnetic field and the size of the magnetic moment. This process corresponds perfectly with the way in which the two basis states $|0\rangle$ and $|1\rangle$ pick up a relative phase shift at the Larmor frequency; this correspondence is not coincidental, but can be seen as a consequence of Ehrenfest's theorem.

The effect of resonant RF fields can be treated in much the same way. The oscillating magnetic field component is divided into two counter-rotating components in the xy plane, one of which rotates around the field in the same direction and at the same rate as the spin. If we transform into a rotating frame which also goes round in the same way then both the magnetization and the RF field component will appear to be static, with the exact position of the RF in the xy plane depending on its phase. The situation now looks just like a magnetic moment in a normal magnetic field, and the spin will precess around the RF field component (which is along, say, the x axis) at a rate which depends on its strength. It is also clear why the counter-rotating component can be ignored: this is moving much faster than the response time of the spin, and so the spin sees it as a rapidly fluctuating field which almost entirely cancels out.

One might ask what has happened to the main magnetic field in this picture. The long answer is that the rotating-frame transformation is an example of a gauge transformation, which results in a gauge field, in this case a fictitious magnetic field that exactly cancels the main field. The short answer is that since the spin does not precess around the field direction in the rotating frame, then the field cannot be there!

The vector model can also be used to model off-resonance excitation. In this case the frame rotates at the RF frequency, not the Larmor frequency, and so the spin is not quite static. This means that the fictitious field does not quite cancel the main field, and a small residual magnetic field remains. The total field experienced by the spin is then the vector sum of the residual field along z and the excitation field along x, and the spin precesses around this vector sum. This sum is longer than the excitation field, and so the precession frequency (Rabi frequency) is increased, but it is tilted away from the xy plane, so that precession will no longer drive the spin from the $+z$ to the $-z$ axis (from $|0\rangle$ to $|1\rangle$).

Finally, as always, it is important to remember that the vector model is not peculiar to the description of nuclear spins, although that is where it is most frequently used. The underlying nature of any two-level quantum system interacting with a classical radiation field is basically the same, and so all these ideas can equally well be applied to atoms in laser fields. This approach ultimately leads to the *optical Bloch equations*, which are analogous to the Bloch equations used to describe the vector model in NMR systems.

3.5 Spin echoes

Spin echoes occur when a 180° rotation (a NOT gate) is placed half-way through a period of evolution under a background Hamiltonian. For a single spin the background Hamiltonian is proportional to σ_z, and so the background evolution will take the form ϕ_z, where the rotation angle ϕ depends on the Larmor frequency, the evolution time, and the choice of frame in which the spin is observed. The total evolution can then be modeled as a quantum network, as discussed in Section 1.4. It takes the form

$$-\boxed{\phi_z}-\boxed{X}-\boxed{\phi_z}- \tag{3.24}$$

and the usual calculations lead to the identity $\phi_z\,180_x\,\phi_z \equiv 180_x$. By this means evolution under the background Hamiltonian can be canceled, making the final state independent of the spin's particular Larmor frequency. Spin echoes are perhaps best known in the context of NMR, but are a universal quantum phenomenon. Some more examples are explored in the exercises.

It is also interesting to look at spin echoes using the vector model. Suppose we have a spin which starts along the x axis, and we allow it to precess at its own Larmor frequency

ω for a time t; this will cause it to rotate in the xy plane through an angle $\phi = \omega t$. Next we apply a NOT gate, which is a $180°$ rotation around the x axis. This leaves the spin within the xy plane, but moves it to a position with angle $-\phi$. After precessing at the same rate ω for another period t the spin is once more found along the x axis, and so the overall effect is that the spin ends the sequence precisely where it started.

The same analysis can be applied to an ensemble of spins which are inhomogeneously broadened, so that different spins have different frequencies. In this case free evolution at the Larmor frequency will cause the spins to precess at different rates, and their individual Bloch vectors will achieve different precession angles in the xy plane. After a long time τ these vectors will be completely scrambled, but the application of a $180°$ rotation followed by further precession will cause the vectors to unscramble. After a further time τ all the spin vectors have come back together, forming an echo of the original state.

Spin echoes are relatively straightforward for single isolated spins, but become more complex in systems with two or more spins which interact with one another. In such cases the $180°$ rotation can be applied to just one spin, or to several spins simultaneously, leading to quite different outcomes. This will be explored in Part II.

3.6 Measurement and initialization

Measurement and initialization are extremely challenging problems in NMR quantum information processing, as a consequence of the low energy scale of NMR transitions (a 1 GHz transition corresponds to an energy of only about 4 μeV). Cooling a nuclear spin into its ground state is, therefore, entirely impractical in the liquid state where most NMR experiments are performed (it is more practical in the solid state, especially if electron spins are used instead). Detecting a single photon corresponding to an NMR transition is clearly extremely challenging, and the very long lifetimes of excited states against spontaneous emission means that detecting fluorescence is completely impractical. Instead, NMR experiments use large ensembles of spins to magnify the signal, and detecting this signal is a largely classical process. The implications of this for NMR quantum computing are explored in Part II.

Further reading

The definitive textbook on NMR, Ernst *et al.* (1987), is rather complex, but many simpler texts are available, such as Goldman (1988), Hore (1995), Freeman (1998), Hore *et al.* (2000), and Levitt (2008). These texts are largely concerned with applications of NMR in the liquid state, which is where most current NMR implementations of quantum information processing can be found, but there are also textbooks on NMR in the solid state, such as Abragam (1983) and Slichter (1989), and on electron spin resonance (ESR), such as Schweiger and Jeschke (2001).

Exercises

3.1 A typical modern NMR spectrometer has a main magnetic field strength of about 12 T, resulting in a ^{1}H Larmor frequency of about 500 MHz, while an RF pulse causing a 90° rotation will typically last around 6 μs. Calculate the strength of the *oscillating* magnetic field component of the RF field.

3.2 Calculate the energy gap between the two spin states of a ^{1}H in the system discussed above. Assuming a Boltzmann distribution between the two energy states, what are the probabilities of finding a given nucleus in the two states at a temperature of 300 K?

3.3 Suppose an NMR sample contains 0.2 ml of water at 300 K: what is the excess number of spins in the lower energy state? What temperature is required to place 99% of the spins in the lower energy state?

3.4 As implied above, a typical NMR sample is a moderately large object (several millimeters in each direction), containing many identical copies of the same spin. If the magnetic field is different at each spin then the Larmor frequency will also vary, giving rise to *inhomogeneous broadening*. Suppose the natural NMR linewidth is around 1 Hz, which is reasonable: how much variation in the field can we tolerate? Is this practical?

3.5 There are many different sequences which can be classified as spin echoes, differing only in fine details. Confirm that $\phi_z\,180_x\,\phi_z \equiv 180_x$, and show that $\phi_z\,180_x\,2\phi_z\,180_x\,\phi_z$ is equivalent to the identity. Similarly show that $\phi_z\,180_x\,\phi_z\,180_x$ and $180_x\,\phi_z\,180_x\,\phi_z$ are also equivalent to the identity. What about $180_y\,\phi_z\,180_y\,\phi_z$ and $180_x\,\phi_z\,180_y\,\phi_z$?

4 Photon techniques

Next we discuss methods for realizing quantum information processing with photons. The approach used here differs greatly from that used for atoms and spins, in that the qubit is not mapped onto two distinct energy levels, but rather onto spatial or polarization degrees of freedom. (Note that to represent a qubit it is only necessary to use two orthogonal states; they do not need to have different energies.)

In *spatial mode encoding*, two orthogonal direction[1] or momentum modes a and b are chosen to represent the qubit states $|0\rangle$ and $|1\rangle$, while in *polarization encoding* the qubit is encoded in the photon polarization, for example using $|H\rangle$ (horizontal polarization) to represent $|0\rangle$ and $|V\rangle$ (vertical polarization) to represent $|1\rangle$. In both cases single-qubit gates can be implemented by linear optical elements, and both forms are commonly used in current experiments. Indeed, since every photon has both spatial and polarization degrees of freedom, many experiments are based on *hybrid encoding*, using the two encoding methods sequentially, converting between them when necessary, or even using both approaches simultaneously.

4.1 Spatial encoding

If two spatial paths impinge on a beam splitter (BS) then this realizes a single-qubit gate by combining the amplitudes in the two paths. For instance, a simple 50/50 beam splitter maps an input state

$$|\psi\rangle_{\text{in}} = \alpha|0\rangle_{\text{in}} + \beta|1\rangle_{\text{in}} \tag{4.1}$$

onto the output state

$$|\psi\rangle_{\text{out}} = \frac{\alpha + \beta}{\sqrt{2}}|0\rangle_{\text{out}} + \frac{\alpha - \beta}{\sqrt{2}}|1\rangle_{\text{out}} = \text{H}|\psi\rangle_{\text{in}}, \tag{4.2}$$

which realizes a Hadamard gate. For general beam splitters the transformation is described by the matrix

$$\text{BS}(\xi, \phi) = \begin{pmatrix} \cos\xi & e^{i\phi}\sin\xi \\ e^{-i\phi}\sin\xi & -\cos\xi \end{pmatrix}, \tag{4.3}$$

---

[1] Note that it is the quantum states represented by the modes which must be orthogonal, not the directions of the two modes: any two paths which are sufficiently well localized to be completely distinct can be used as orthogonal modes. In drawings of interferometers the states $|0\rangle$ and $|1\rangle$ will normally correspond to the upper and lower arms, respectively.

where $\cos^2 \xi$ and $\sin^2 \xi$ are the reflectivity and transmittivity of the beam splitter, respectively. The simple 50/50 beam splitter then corresponds to $\mathrm{BS}(\pi/4, 0) = \mathrm{H}$. The value of ϕ is the phase shift experienced on transmission through the beam splitter, and is similar to the phase gate discussed next.

A single-qubit phase gate Φ is implemented by putting a slab of transparent medium with refractive index n_1 and length L into the path of one spatial mode, say $|1\rangle$. This causes a phase shift $\phi = (n_1 - n_0)L\omega/c$ with respect to the second arm, which we assume to be in air where the refractive index is n_0, and so can be described as a *phase shifter*. This maps the qubit wave function according to the truth table

$$|0\rangle_{\mathrm{in}} \rightarrow |0\rangle_{\mathrm{out}} \qquad |1\rangle_{\mathrm{in}} \rightarrow e^{i\phi}|1\rangle_{\mathrm{out}}, \qquad (4.4)$$

and choosing $\phi = \pi/4$ gives the phase gate T. As the logic gates T and H are universal for single-qubit logic gates, these two linear optical elements allow the realization of all single-qubit operations.

Example 4.1 We can interpret the familiar setup of a Mach–Zehnder interferometer in terms of single-qubit gates. The interferometer consists of a 50/50 beam splitter, followed by a phase shifter Φ in one arm, which we will assume to correspond to $|1\rangle$, and then a second 50/50 beam splitter. As described below, we assume that the two arms of the interferometer have been carefully adjusted to have the same length. The interferometer then maps the input state $|\psi\rangle_{\mathrm{in}}$ onto

$$|\psi\rangle_{\mathrm{out}} = \mathrm{H}\Phi\mathrm{H}|\psi\rangle_{\mathrm{in}}, \qquad (4.5)$$

which in matrix form is given by

$$|\psi\rangle_{\mathrm{out}} = \frac{1}{2} \begin{pmatrix} 1 & 1 \\ 1 & -1 \end{pmatrix} \begin{pmatrix} 1 & 0 \\ 0 & e^{i\phi} \end{pmatrix} \begin{pmatrix} 1 & 1 \\ 1 & -1 \end{pmatrix} \begin{pmatrix} \alpha \\ \beta \end{pmatrix} \qquad (4.6)$$

$$= e^{i\phi/2} \begin{pmatrix} \cos(\phi/2)\alpha - i\sin(\phi/2)\beta \\ -i\sin(\phi/2)\alpha + \cos(\phi/2)\beta \end{pmatrix}. \qquad (4.7)$$

If the system starts in $|0\rangle$ and the output of the interferometer is measured in the usual basis, then the result is $|0\rangle$ with probability $\cos^2(\phi/2)$ and $|1\rangle$ with probability $\sin^2(\phi/2)$.

Note that this brute force calculation is, however, unnecessary. The phase shift gate can be rewritten as

$$\Phi = e^{i\phi/2} \begin{pmatrix} e^{-i\phi/2} & 0 \\ 0 & e^{i\phi/2} \end{pmatrix} \qquad (4.8)$$

and so is equivalent (up to an irrelevant global phase) to a ϕ_z gate. Thus the Mach–Zehnder network is essentially identical to the network for Ramsey fringes explored in Section 2.7, and the outcome of the experiment is already known. This equivalence is hardly surprising, as it has already been noted that Ramsey fringes are equivalent to an interference pattern, but this is an example of the general phenomenon that apparently disparate experiments are seen to be the same when viewed from the standpoint of quantum information theory.

4.2 Polarization encoding

In polarization encoding, single-qubit gates are implemented using birefringent wave plates, which can act as both polarization rotators and polarization phase shifters. Birefringent materials are optically anisotropic, and the refractive index depends on the direction of propagation and the polarization of the light wave with respect to the crystal axes. The simplest case occurs for *uniaxial* materials such as calcite, where the optical properties are different along one of the three axes, called the *extraordinary axis*. If light propagates along this direction, also called the *optic axis*, then the refractive index is independent of polarization, and no special behavior is observed. Suppose, however, that the crystal is cut so that the extraordinary axis is parallel to the crystal surface. Light propagating through the crystal, along one of the *ordinary axes*, can be polarized either along the other ordinary axis or along the extraordinary axis, and the refractive index will depend on the choice of polarization.

These two axes can also be called the *fast* and *slow* axes, but the fast axis can correspond to either the ordinary or the extraordinary axis, depending on the material. Suppose the crystal is arranged so that the fast axis is aligned with horizontally polarized light, representing $|0\rangle$, while the slow axis is aligned with vertical light, representing $|1\rangle$. Propagating through the crystal then applies an additional phase shift ϕ to state $|1\rangle$, implementing the phase gate Φ, and choosing the thickness of the crystal so that the difference in optical path lengths for the two polarizations is $\lambda/4$ gives a *quarter wave plate*, which implements the traditional phase gate S. Similarly a *half wave plate* aligned in this way will implement Z.

The behavior is more complex if the fast and slow axes are not aligned with the polarization basis states, and this can give rise to rotations of the polarization. Suppose the crystal is now arranged with the fast axis aligned at an angle θ from the horizontal axis toward the vertical axis. The eigenstates of the phase gate, that is the states which only pick up phases as described above, are

$$|f\rangle = \cos\theta|0\rangle + \sin\theta|1\rangle \quad \text{and} \quad |s\rangle = \cos\theta|1\rangle - \sin\theta|0\rangle \qquad (4.9)$$

and the basis states can now be decomposed in terms of these eigenstates

$$|0\rangle = \cos\theta|f\rangle - \sin\theta|s\rangle \quad \text{and} \quad |1\rangle = \cos\theta|s\rangle + \sin\theta|f\rangle. \qquad (4.10)$$

Thus the action of the crystal is

$$|0\rangle \rightarrow \cos\theta|f\rangle - \sin\theta\,e^{i\phi}|s\rangle = (\cos^2\theta + \sin^2\theta e^{i\phi})|0\rangle + \cos\theta\sin\theta(1 - e^{i\phi})|1\rangle, \quad (4.11)$$

$$|1\rangle \rightarrow \cos\theta\,e^{i\phi}|s\rangle + \sin\theta|f\rangle = (\cos^2\theta e^{i\phi} + \sin^2\theta)|1\rangle + \cos\theta\sin\theta(1 - e^{i\phi})|0\rangle, \quad (4.12)$$

corresponding to the unitary transformation

$$U(\theta, \phi) = \begin{pmatrix} \cos^2\theta + \sin^2\theta e^{i\phi} & \cos\theta\sin\theta(1 - e^{i\phi}) \\ \cos\theta\sin\theta(1 - e^{i\phi}) & \cos^2\theta e^{i\phi} + \sin^2\theta \end{pmatrix} \qquad (4.13)$$

and choosing θ and ϕ allows a wide range of gates to be implemented. In particular, using a quarter wave plate ($\phi = \pi/2$) aligned at 45° to the basis states ($\theta = \pi/4$) gives the operation

$$U(\pi/4, \pi/2) = \frac{1}{2} \begin{pmatrix} 1+i & 1-i \\ 1-i & 1+i \end{pmatrix} = \frac{e^{i\pi/4}}{\sqrt{2}} \begin{pmatrix} 1 & -i \\ -i & 1 \end{pmatrix} \qquad (4.14)$$

which is (up to an irrelevant global phase) a SQUARE-ROOT-OF-NOT gate. This gate can then be combined with S gates to make a Hadamard gate. Alternatively, and more sensibly, other important gates can be implemented directly by choosing appropriate values of θ and ϕ.

4.3 Single-photon sources and detectors

The simplest way to obtain a single-photon source is to use pulsed laser light which is a coherent state of light, that is a superposition $|\alpha\rangle$ of different photon number states $|n\rangle$ given by

$$|\alpha\rangle = e^{-|\alpha|^2/2} \sum_{n=0}^{\infty} \frac{\alpha^n}{\sqrt{n!}} |n\rangle. \qquad (4.15)$$

The first few terms in this sum can be expanded as

$$|\alpha\rangle \propto |0\rangle + \alpha|1\rangle + \frac{\alpha^2}{\sqrt{2}}|2\rangle + \frac{\alpha^3}{\sqrt{6}}|3\rangle + \dots \qquad (4.16)$$

where $|0\rangle$ and $|1\rangle$ are now number states, corresponding to zero photons and one photon respectively, rather than qubit states. By attenuating this beam the laser pulse intensity and thus the parameter $|\alpha|^2$ goes down. Then the dominant contributions to the state of the laser pulse are the vacuum $|0\rangle$ and the single photon state $|1\rangle$, with all higher-order terms decreasing with powers of α. For instance, for the case $\alpha = \sqrt{0.1}$, less than 10% of laser pulses actually contain any light, but these pulses are single photons with around 95% probability.

Qubits in desired initial states can then be prepared by directing the path of these photons (for spatial encoding) or by using a polarizing filter (for polarization encoding). The drawback of this simple method is that the source does not indicate whether a photon is present or not. More sophisticated schemes are currently being developed which allow the on-demand generation of a single photon with well-defined polarization, wavelength and direction. Such single-photon sources are important for improved implementations of quantum communication schemes.

Single-photon detectors required in quantum communication should have a high quantum efficiency, detect photons over a broad frequency range (100 nm to 2000 nm), and have low dark count rates (false counts in the absence of photons). They should also recover quickly after detecting a photon, allowing resolution of close pairs of photons. Single-photon detectors can be further divided into photon number-resolving devices (which allow individual number states to be distinguished) and photon counting devices (which only allow the vacuum state to be distinguished from all other number states).

Photon detectors can directly measure spatially encoded qubits by simply placing a detector in each path. Note that the probability of a click in a given detector is given by the square modulus of the amplitude of the corresponding wave; this is exactly as expected from quantum mechanics, where the probability of obtaining a given measurement result in a superposition state is given by the square modulus of the corresponding amplitude as described in Section 1.5, and is also precisely as expected from classical wave optics, where the intensity of a light wave depends on the square modulus of the wave amplitude. Polarization states can be distinguished by using a polarizing beam splitter (which is made from birefringent materials and directs horizontally and vertically polarized photons along different paths) to convert polarization encoding to spatial encoding.

4.4 Conventions

We will analyze a number of optical setups in Part III, and will use the following conventions to simplify notation and discussion unless stated otherwise:

- For spatially encoded qubits the upper path will be identified with $|0\rangle$ and the lower path with $|1\rangle$, while for polarization encoded qubits we use $|H\rangle$ to represent $|0\rangle$ and $|V\rangle$ to represent $|1\rangle$.
- Beam splitters (indicated by BS) are assumed to implement a Hadamard gate between input and output paths.
- Polarizing beam splitters (indicated by PBS) are assumed to transmit horizontally polarized photons and reflect vertically polarized photons, but a rotation by an angle ϕ from this standard setting may also be indicated.
- The paths of an interferometer are assumed to have the same optical path length; devices for adjusting the path lengths so that this is true will not be shown.
- The reduction in spatial overlap caused by a phase shifter ϕ is assumed to be negligible.
- A delay line drawn in one arm of an interferometer is assumed to delay the photon pulse by more than its coherence length so that *no* interference between the two arms is possible for photons which entered the system at the same time.

These conventional definitions allow us to ignore many experimental details, which are essential in practice and which can make real optics experiments quite challenging. This follows our approach throughout this book of concentrating on the essential properties of qubits, rather than on detailed matters of implementation.

Further reading

Many textbooks say relatively little about elementary quantum information processing with photons, often starting directly with entangled two-qubit states. However, the basic operations are well described in standard optics texts such as Hecht (2002). Note that the quantum mechanical description of polarized photons is essentially identical to the Jones vector description of fully polarized light.

Exercises

4.1   We have already explored some of the properties of birefringent wave plates, with a particular emphasis on quarter wave plates ($\phi = \pi/2$). Now evaluate the unitary transformation performed by a half wave plate, and show how this can be used to implement NOT gates and Hadamard gates directly.

4.2   Show that the coherent state $|\alpha\rangle$ is correctly normalized. Find, as a function of $|\alpha|$, the fraction of laser pulses containing at least one photon, and the fraction of such pulses containing exactly one photon. Hence confirm the results for $\alpha = \sqrt{0.1}$ given in the text.

5 Two qubits and beyond

As we have seen, even a single qubit is a surprisingly interesting object. However, the real power of quantum information processing begins with systems of two or more qubits. Before studying these in detail we need to expand our notation a little.

Consider a system of two qubits, labeled a and b, each of which has two basis states, $|0\rangle$ and $|1\rangle$. The whole system then has four basis states, which can be written as $|0_a 0_b\rangle$, $|0_a 1_b\rangle$, $|1_a 0_b\rangle$ and $|1_a 1_b\rangle$, and can be found in any general superposition of these states, so that it occupies a four-dimensional Hilbert space. In the same way, a system of three qubits inhabits an eight-dimensional Hilbert space, and so on. This exponential increase in the size of the Hilbert space with a linear increase in the number of qubits underlies the power of quantum computers.

5.1 Direct products

The size of the Hilbert spaces involved can also be a huge problem, however, making it difficult to describe states of systems with many qubits. A partial solution is to note that some states can be described in a simpler way, using the concept of *direct products*. These states, in which the individual qubits can in principle be discussed separately, make up a tiny minority of the states accessible to a multi-qubit system, but include many important states, most notably the basis states. States of this kind are said to be *separable*, and states which are not separable are said to be *entangled*. Entangled states are much more interesting than separable ones, but it is wise to begin with the simpler case.

By the basis state $|1_a 0_b\rangle$ we mean a state where qubit a is in state $|1\rangle$ and qubit b is in state $|0\rangle$, and we can write this as $|1_a\rangle \otimes |0_b\rangle$, where the symbol $\otimes$ indicates a direct product. For the moment we shall not worry too much about what a direct product really is, and just think of it as a way of combining two subsystems; a more mathematical discussion can be found in the next section. There is considerable variation in the way these states are described: $|1_a 0_b\rangle$ can also be written as $|1\rangle \otimes |0\rangle$, as $|10\rangle$, or most simply of all as $|2\rangle$, where this last version is obtained by interpreting the 1 and the 0 as the two bits making up the binary number 10, or decimal 2. Most authors move back and forth between these different notations, using whatever is most convenient at the time. Of course, if the more compact forms are used, it is essential to use a consistent ordering of the qubits to avoid ambiguous notation.

The direct product approach can also be used to describe more complex states. Suppose, for example, a Hadamard gate is applied to the second qubit of a system starting in the state

$|00\rangle$. This can be written as

$$|00\rangle = |0\rangle \otimes |0\rangle \xrightarrow{H_b} |0\rangle \otimes |+\rangle = |0\rangle \otimes (|0\rangle + |1\rangle)/\sqrt{2} = (|00\rangle + |01\rangle)/\sqrt{2}. \quad (5.1)$$

Similarly, direct products can be used to write down single-qubit operators in a multi-qubit system without the need for explicit labels: thus, for example, $H_b = \mathbb{1} \otimes H$ (that is, do nothing to the first qubit and apply a Hadamard gate to the second qubit), while $H_a = H \otimes \mathbb{1}$. Simultaneous Hadamard gates can also be applied to both qubits, using $H^{\otimes 2} = H \otimes H$. Gates of this kind, where the same operation is applied to both members of a pair of qubits, are sometimes called *bilateral* operations.

5.2 Matrix forms

Much of the point of the direct product approach is to avoid writing out explicit matrix descriptions of states of multi-qubit systems, but sometimes it is useful to do so. The basic idea behind a direct product is to multiply a copy of the second matrix by each element of the first matrix in turn: thus

$$\begin{pmatrix} a \\ b \end{pmatrix} \otimes \begin{pmatrix} \alpha \\ \beta \end{pmatrix} = \begin{pmatrix} a\alpha \\ a\beta \\ b\alpha \\ b\beta \end{pmatrix}. \quad (5.2)$$

Note that, for example, the matrix representation of $|10\rangle$ is

$$\begin{pmatrix} 0 \\ 1 \end{pmatrix} \otimes \begin{pmatrix} 1 \\ 0 \end{pmatrix} = \begin{pmatrix} 0 \\ 0 \\ 1 \\ 0 \end{pmatrix}, \quad (5.3)$$

exactly what would be naively expected. An equivalent approach can be used for operators, so that the matrix representation of H_b is

$$\begin{pmatrix} 1 & 0 \\ 0 & 1 \end{pmatrix} \otimes \frac{1}{\sqrt{2}} \begin{pmatrix} 1 & 1 \\ 1 & -1 \end{pmatrix} = \frac{1}{\sqrt{2}} \begin{pmatrix} 1 & 1 & 0 & 0 \\ 1 & -1 & 0 & 0 \\ 0 & 0 & 1 & 1 \\ 0 & 0 & 1 & -1 \end{pmatrix}. \quad (5.4)$$

When an operator separately affects two different qubits it may be useful to use the fact that the operator can be considered as two sequential operators, one affecting each qubit; thus applying $H^{\otimes 2}$ is the same as applying H_a followed by H_b, or *vice versa*. Similarly direct products and conventional matrix products can be carried out in either order,

$$(a \otimes b) \cdot (c \otimes d) = (a \cdot c) \otimes (b \cdot d). \quad (5.5)$$

These methods frequently allow calculations to be simplified.

5.3 Two-qubit gates

We have already seen some two-qubit gates: for example H_b implements a single-qubit Hadamard in a two-qubit system, while $H^{\otimes 2}$ represents simultaneous Hadamard gates in a two-qubit system. However these gates, which can all be written using direct products, are in some sense a trivial extension of the corresponding gates in a single-qubit system, and are usually described as single-qubit gates. A much more interesting two-qubit gate is the controlled-NOT gate, which has the explicit matrix form

$$
\begin{pmatrix}
1 & 0 & 0 & 0 \\
0 & 1 & 0 & 0 \\
0 & 0 & 0 & 1 \\
0 & 0 & 1 & 0
\end{pmatrix},
\tag{5.6}
$$

and a little thought shows that this matrix cannot be written as a direct product. It seems that this operator might be more interesting than those discussed above, and this is indeed the case. In particular, the controlled-NOT gate is a key gate in the generation of entangled states. Furthermore, it can be shown that the combination of the controlled-NOT gate and a small set of single-qubit gates is *universal* for quantum information processing, meaning that any desired operation can be built from a network of these gates.

The reason why this gate is called a controlled-NOT gate can easily be seen by applying it to the four basis states in turn, effectively evaluating its *truth table*. Clearly $|00\rangle$ and $|01\rangle$ are unaffected, while $|10\rangle$ and $|11\rangle$ are interchanged. Thus, the effect of the controlled-NOT gate is to apply a NOT gate to the second qubit *if and only if* the first qubit is in state $|1\rangle$. This is an example of controlled evolution, in which the state of one qubit is used to influence the state of another, a process at the heart of computation.

Yet another way of looking at the action of the controlled-NOT gate is to use the concept of *bitwise addition modulo 2*, which simply means adding two bits, throwing away any carries that are generated. Thus $0 \oplus 0 = 0$ and $0 \oplus 1 = 1 \oplus 0 = 1$ as normal, but $1 \oplus 1 = 0$ as the carry is simply ignored. Note that $a \oplus b$ is equal to zero if a and b are the same, and is equal to one if a and b are different. Alternatively, $a \oplus b$ is equal to the XOR (exclusive-OR) of a and b.

A final description of the controlled-NOT gate is provided by noting that NOT is equivalent to the X gate, and then seeking an expansion of the gate in terms of direct products. The action of controlled-NOT is to apply $\mathbb{1}$ to qubit 2 if qubit 1 is in state $|0\rangle$, and to apply X to qubit 2 if qubit 1 is in state $|1\rangle$. This can be written as

$$
\text{controlled-NOT} = |0\rangle\langle 0| \otimes \mathbb{1} + |1\rangle\langle 1| \otimes \sigma_x,
\tag{5.7}
$$

which can be confirmed by direct matrix calculations; if desired, the term $|0\rangle\langle 0|$ can be replaced by $\frac{1}{2}(\mathbb{1} + \sigma_z)$, while $|1\rangle\langle 1|$ can be replaced by $\frac{1}{2}(\mathbb{1} - \sigma_z)$. Note that in cases such as this, where an operator is written as a *sum* of direct products, it is essential to be careful about global phases: it is not possible to simply replace the σ_x in equation (5.7) by 180_x as these differ by a factor of i.

The controlled-NOT gate is commonly used in theoretical discussions of quantum information processing, but in many experimental implementations it is easier to use a closely related gate, the controlled-Z gate, which performs the transformation

$$|11\rangle \xrightarrow{\text{c-Z}} -|11\rangle \tag{5.8}$$

while leaving the other three basis states unaffected. This can be converted to a controlled-NOT gate using a pair of Hadamard gates; the equivalence can easily be proved by brute force multiplication

$$\frac{1}{\sqrt{2}}\begin{pmatrix} 1 & 1 & 0 & 0 \\ 1 & -1 & 0 & 0 \\ 0 & 0 & 1 & 1 \\ 0 & 0 & 1 & -1 \end{pmatrix}\begin{pmatrix} 1 & 0 & 0 & 0 \\ 0 & 1 & 0 & 0 \\ 0 & 0 & 1 & 0 \\ 0 & 0 & 0 & -1 \end{pmatrix}\frac{1}{\sqrt{2}}\begin{pmatrix} 1 & 1 & 0 & 0 \\ 1 & -1 & 0 & 0 \\ 0 & 0 & 1 & 1 \\ 0 & 0 & 1 & -1 \end{pmatrix} = \begin{pmatrix} 1 & 0 & 0 & 0 \\ 0 & 1 & 0 & 0 \\ 0 & 0 & 0 & 1 \\ 0 & 0 & 1 & 0 \end{pmatrix} \tag{5.9}$$

or by decomposing the controlled-Z gate as a sum of direct products and then applying the approach of equation (5.5) to each term in the sum.

5.4 Networks and circuits

Unlike networks of single-qubit gates, networks of two-qubit gates are usually difficult to describe in simple words, and it is much better to draw a *quantum circuit*. For example, equation (5.9) can be redrawn as a network, as shown below:

$$\tag{5.10}$$

In circuits like this each horizontal line or "wire" corresponds to one qubit, and time runs from left to right, so that the leftmost gate is the first gate applied. Single-qubit gates are drawn on the relevant line, while two-qubit gates connect two lines.

The gate in the middle of this equation is a controlled-X gate, and the symbol is made up of three parts. On the top line is a small filled circle, indicating that this qubit *controls* the gate. This control mark is connected by a vertical line to the X gate on the second qubit (the *target* qubit). As the X gate is equivalent to a NOT operation, this is also a controlled-NOT gate. The controlled-NOT gate is very common in quantum networks, and is more usually written as shown on the far right, with the symbol ⊕ indicating a NOT gate (this is a slight abuse of notation, which can partly be justified by noting that $a \oplus 1 = \text{NOT}(a)$). Note that in larger systems a control line may need to cross one or more wires corresponding to other qubits; if the control line is not connected to a wire by a control circle, then these intervening qubits play no role in the gate's operation.

On the left-hand side we have three gates, including two single-qubit Hadamard gates, applied to the second qubit, and a peculiar two-qubit gate, comprising two control dots connected by a control line. Once again this is an example of abuse of notation, and is used to indicate a controlled-Z gate,

$$\text{(circuit diagram)} \qquad (5.11)$$

The justification for this abuse is the fact that the controlled-Z gate, unlike the controlled-NOT gate, is *symmetric* between the control and target qubits: it is not meaningful to say which is which, as the nominal roles could be interchanged with no effect. This symmetry is a characteristic of physical interactions, and explains the importance of the controlled-Z gate in physical implementations.

An interesting circuit which can be built entirely out of controlled-NOT gates is the SWAP circuit shown below:

$$\text{(circuit diagram)} \qquad (5.12)$$

which acts to interchange the states of two different qubits, effectively making two quantum wires cross over. Exploring this circuit is a good first exercise in playing with gates.

5.5 Entangled states

We have already hinted at the existence of entangled states, which are states of a system of two or more qubits which cannot be written as a direct product of single-qubit states. Here we will confine ourselves to two-qubit systems where the phenomenon of entanglement is relatively simple and well understood. A simple way to generate an entangled state from a basis state is to use the network

$$\text{(circuit diagram)} \qquad (5.13)$$

and we can follow through this network a step at a time:

$$|00\rangle \xrightarrow{H_a} (|00\rangle + |10\rangle)/\sqrt{2} \qquad (5.14)$$

$$\xrightarrow{c\text{-}X} (|00\rangle + |11\rangle)/\sqrt{2}, \qquad (5.15)$$

where the first line follows from the properties of the Hadamard gate and the second line is obtained by using the fact that the controlled-NOT gate is a unitary operation, and thus a linear operation, and so its effect on a superposition can be obtained by applying the gate to each of the terms in turn. The final state looks simple enough, but has some very peculiar properties! This state is *inseparable*, which means that it cannot be written as a direct product of states of the two individual qubits. In turn this means that the properties of the state cannot fully be described by listing the properties of the two qubits involved: rather, they are properties of the two-qubit state taken as a whole.

To take a simple example, consider measuring the state of the first qubit. The system is in an equally weighted superposition of two states, in one of which the first qubit is in $|0\rangle$, and in the other the first qubit is in $|1\rangle$. Thus any measurement of the first qubit will return either $|0\rangle$ or $|1\rangle$, at random and with equal probability. This is not particularly odd; what is more odd is the effect that measuring the first qubit has on the *second* qubit. Our entangled state is a superposition of two states, in both of which the two individual qubits have the same state; thus if we measure qubit a and find it is in $|0\rangle$ we know immediately that qubit b must also be in state $|0\rangle$. Similarly, if we find qubit a in $|1\rangle$ then qubit b will also be in $|1\rangle$. The behavior of the two qubits is completely intertwined, or *entangled*.[1]

The state discussed above is only one example of an infinite number of possible entangled states. Particularly important among these are the four Bell states, which are *maximally entangled* states, meaning that their behavior is as unlike a direct product state as possible. These states are defined by

$$|\phi^{\pm}\rangle = (|00\rangle \pm |11\rangle)/\sqrt{2} \qquad |\psi^{\pm}\rangle = (|01\rangle \pm |10\rangle)/\sqrt{2}, \qquad (5.16)$$

and so the state discussed previously is a $|\phi^{+}\rangle$ Bell state. Note that the four maximally entangled Bell states form an orthonormal basis for two-qubit states. Of course, general superpositions of the four Bell basis states need not be entangled.

Further reading

Discussions of universal sets of quantum gates and methods for constructing quantum networks can be found in standard texts such as Nielsen and Chuang (2000), Mermin (2007), and Stolze and Suter (2008). There are some useful examples and exercises on elementary quantum circuits in Schumacher and Westmoreland (2010).

Exercises

5.1   Show that a controlled-NOT gate can be built out of Hadamard gates and a controlled-σ_z gate without using explicit matrices in your argument.

5.2   Use the "bitwise addition modulo 2" description of the controlled-NOT gate to show that a network of three controlled-NOT gates will swap the values of two qubits in

---

[1] The behavior described above does not in fact prove that the two qubits are entangled, as the same results can be obtained using classically correlated states. Entangled states, however, can show correlations over a range of possible measurements which exceed the limits possible for any classically correlated state. This point will be explored briefly in the exercises for the next chapter, and will be discussed in considerable detail in Chapter 16.

eigenstates. Hence show that this network acts as a SWAP gate for any separable state of two qubits.

5.3 Calculate an explicit matrix form for the SWAP gate. What does this gate do to a pair of qubits in a Bell state? Why is this answer not surprising?

5.4 Find explicit expressions for the four computational basis states of a two-qubit system in terms of superpositions of the four Bell states.

5.5 Show that the entangling network shown in equation (5.13) can be used to produce all four Bell states by using different initial states for the input qubits.

5.6 How can the four Bell states be converted into four distinguishable states in the computational basis?

Measurement and entanglement

We now have all the basic tools we need to describe systems of one and two qubits. In this final chapter of Part I we look at some of the consequences of the peculiar properties of qubits when they are used to encode information. Many of these results can be traced to the properties of quantum measurement.

6.1 Measuring a single qubit

A key result in quantum information theory is that it is impossible to accurately characterize a single qubit. In other words, there is no experiment, or sequence of experiments, which allows us to find out the state of a single quantum bit.

The reason for this problem is twofold. Firstly, we have to make some sort of decision about the basis we will use for the measurement. For example, when measuring a single qubit the most popular choice is to make a measurement in the computational basis. This is equivalent to asking the qubit whether it is in state $|0\rangle$ or state $|1\rangle$. If the qubit is indeed in one of the basis states the measurement process is simple, and we will get the obvious answer of 0 if it is in $|0\rangle$ and 1 if it is in $|1\rangle$. If, however, the qubit is in a superposition, such as $\alpha|0\rangle + \beta|1\rangle$, then the situation is more difficult. Characterizing the state now means determining the values of the two complex numbers α and β, but the measurement can only return the answer 0 or 1, and for a superposition state one of these two answers will be returned at random, with probabilities $|\alpha|^2$ and $|\beta|^2$, respectively.

Clearly we cannot characterize a superposition state in a single measurement, but why not just make repeated measurements, and so gain statistical information about α and β? The problem is that the first measurement does not leave the state unaffected: if the first measurement returned 0 then the state is *changed* to $|0\rangle$, and if the first measurement returned 1 the state is changed to $|1\rangle$. Any subsequent measurement of the state will therefore return the same answer as before, and no more information can be obtained. This point was addressed briefly in Section 1.5.

One might wonder what is so special about the computational basis, and the short answer is that measuring in any other orthonormal basis is ultimately equivalent. The key thing about the computational basis is simply that it is an orthonormal basis, as the only quantum measurements that work reliably are those where the possible answers correspond to orthonormal states. However, any orthonormal basis is (in principle) as good as any other, and we could choose some other basis without altering the underlying ideas.

If we do know something about the state of the qubit, the situation is changed. For example, if we know the qubit is in either $|0\rangle$ or $|1\rangle$ then a measurement in the computational basis will immediately tell us which one it is. Next consider a qubit which is known to be in one of the two superposition states $|\pm\rangle = (|0\rangle \pm |1\rangle)/\sqrt{2}$. In each case measurement in the computational basis will return either $|0\rangle$ or $|1\rangle$ with 50% probability, and so this measurement tells us nothing useful about the state. The two states are, however, orthonormal, and so should be completely distinguishable. This can be achieved by measuring in the X-basis, which always returns either $|+\rangle$ or $|-\rangle$.

Example 6.1 Consider the effect of the network

$$-\boxed{H}-\boxed{\measuredangle}-\boxed{H}- \tag{6.1}$$

applied to a qubit in a general state, which we can write in the X-basis as $|\psi_x\rangle = \alpha|+\rangle+\beta|-\rangle$. The first Hadamard gate will convert this to the corresponding state in the Z-basis, $|\psi_z\rangle = \alpha|0\rangle + \beta|1\rangle$, and the conventional Z-measurement gate will project it onto either $|0\rangle$ or $|1\rangle$, with probabilities $|\alpha|^2$ and $|\beta|^2$, respectively. Finally, the second Hadamard gate will rotate these basis states back onto $|+\rangle$ and $|-\rangle$. Thus the whole process can be written as

$$\alpha|+\rangle + \beta|-\rangle \longrightarrow |\alpha|^2|+\rangle\langle+| + |\beta|^2|-\rangle\langle-|, \tag{6.2}$$

which is a measurement in the X-basis. Measurements in any other orthonormal basis can be achieved in a similar way.

It is also possible to consider making measurements that seek to distinguish states that are not orthonormal; the detailed answers are quite complicated, but the key result is that such measurements cannot be made to work with perfect reliability, and including them does not fundamentally change the discussion above.

Now suppose we have a qubit in either $|0\rangle$ or $|1\rangle$ and choose to measure its state in the X-basis. In this case the measurement will return either $|+\rangle$ or $|-\rangle$, each with 50% probability, and we learn nothing at all about the state. While we can optimize our measurement process for any particular pair of states, we cannot simultaneously optimize it for all possible states. Clearly, if the state of the qubit is unknown it is impossible to optimize the measurement.

More insight into the measurement process can be obtained by using a Bloch sphere picture. The two states $|0\rangle$ and $|1\rangle$ lie at the north and south poles of the Bloch sphere, and a measurement in the computational basis applied to a general state is essentially a method of estimating its projection onto the z axis. The two basis states $|0\rangle$ and $|1\rangle$ lie at opposite ends of the z axis, and are easily distinguished. By contrast, the states $|\pm\rangle$ lie at opposite ends of the x axis, and the projection onto the z axis is zero for both states, showing that they cannot be distinguished. These states are best distinguished by their projections onto the x axis, that is by measurements in the X-basis, but this is completely useless for the states $|0\rangle$ and $|1\rangle$, whose projection onto the x axis is zero. A measurement will only be perfect if the measurement axis is parallel to the state, and will be completely useless if the measurement axis is perpendicular to the state. For a completely unknown state there is no sensible way to choose the axis, and any measurement is as good as any other.

The classic example of a quantum measurement is a Stern–Gerlach apparatus, which measures the projection of a spin onto some axis. For a spin-$\frac{1}{2}$ particle this is, of course, entirely equivalent to measuring a qubit in some basis: a Stern–Gerlach apparatus aligned along the z axis is equivalent to a measurement in the computational basis, while one aligned along the x axis is equivalent to a measurement in the X-basis, aligned with the x axis of the Bloch sphere. These measurements always produce one of two results: the Bloch vector corresponding to the spin state is either parallel or antiparallel to the measurement axis. If the spin is neither parallel nor antiparallel then one of the two permitted results is returned at random, with probabilities depending on the projection of the spin onto the axis.

The situation becomes much more interesting when considering a cascaded network of measurements, such as that shown below:

$$
|\psi\rangle - \boxed{Z} \begin{matrix} - |0\rangle \\ - |1\rangle \end{matrix} \quad |0\rangle - \boxed{X} \begin{matrix} - |+\rangle \\ - |-\rangle \end{matrix} \quad - \boxed{Z} \begin{matrix} - |0\rangle \\ - |1\rangle \end{matrix} \tag{6.3}
$$

Suppose we measure the z-component of a spin, and find it to be $|0\rangle$, and then measure the x-component of a spin, and find it to be $|+\rangle$. Finally we measure the z-component again, and find that the result is either $|0\rangle$ or $|1\rangle$, each with 50% probability. This is easy to explain using quantum mechanics (the effect of the x-measurement is to *change* the state to either $|+\rangle$ or $|-\rangle$), but hard to make sense of in any other way.

The same result can be demonstrated experimentally even more simply using polarized light. In this case horizontally and vertically polarized light corresponds to the computational basis states, $|0\rangle$ and $|1\rangle$, while light polarized at $\pm45°$ corresponds to the $|\pm\rangle$ states. Suppose a photon is passed through a pair of crossed polarizers: the only photons that can get through are those which were in $|0\rangle$ at the first polarizer and in $|1\rangle$ at the second one, and so no light is transmitted. But if a polarizer at $\pm45°$ is placed *between* the two crossed polarizers, then the intervening measurement changes the states of the photons, allowing some of them to pass!

Example 6.2 Take a beam of vertically polarized light, and pass it through an ideal piece of polaroid film with a vertical axis. The light beam will be completely transmitted. Now place a second polarizer after the first one, at an angle θ; the transmitted fraction will drop to $\cos^2\theta$, with no transmission occurring at 90° (the Law of Malus). Now consider using two ideal polarizers after the first one, at angles of 45° and 90°: what will be the transmitted fraction in this case? Assume perfect polarizers throughout.

Solution

For the case of two polarizers the transmitted fraction is $[\cos(\pi/4)]^4 = 1/4$. Note that there is no loss at the first polarizer as the light is pre-aligned.

Example 6.3 Now consider a sequence of $n + 1$ polarizers, equally spaced up to $90°$ (so that for the case $n = 3$ the first polarizer is at $0°$ and the next three are at $30°$, $60°$ and $90°$, respectively). What is the transmission for general values of n? What is the value in the limit $n \to \infty$?

Solution

The previous result generalizes in the obvious way to $[\cos(\pi/2n)]^{2n}$, and in the infinite limit the transmitted fraction goes to one. Thus a sequence of quantum measurements can act to rotate the quantum state.

The behavior seen above is related to another phenomenon, the *quantum Zeno effect*. Just as a series of changing quantum measurements can rotate a quantum state, a series of fixed quantum measurements can suppress the evolution of a quantum state. This is briefly explored in the exercises.

6.2 Ensembles and the no-cloning theorem

So far we have assumed that we have only one copy of our unknown quantum state. Suppose, however, that we have a large number of identical copies of the state, a situation usually called an ensemble. If the state is $|\psi\rangle$ then the ensemble can be written as

$$|\psi\rangle^{\otimes n} = \overset{n}{\bigotimes} |\psi\rangle = \overbrace{|\psi\rangle \otimes |\psi\rangle \otimes \cdots \otimes |\psi\rangle}^{n \text{ terms}}, \tag{6.4}$$

which is a direct product of copies of the unknown state. The significance of the direct product form is that the individual copies of the state are *independent*, in the sense that manipulating one qubit does not affect any others. By performing several different measurements on many copies of the state we can get a very good idea of what the state is. It might seem that this offers a solution to the problem of characterizing an unknown state: all that is necessary is to make an ensemble of copies of the state and then measure these. As we shall see, however, this process is impossible.

The no-cloning theorem is one of the most important results in the whole of quantum information theory, but it is also one of the simplest. It is possible to copy classical information without limits, but it is almost trivial to prove that an unknown quantum state cannot be copied (cloned), and a brief proof is sketched below.

The proof proceeds by contradiction. Suppose a quantum cloning device, capable of accurately copying a completely unknown state, did in fact exist. Clearly such a device must be capable of copying the two basis states $|0\rangle$ and $|1\rangle$. As the process must be performed reversibly, the cloning device cannot simply conjure new qubits out of nothing, and so the copies must overwrite an additional *ancilla* qubit which starts in some initial state, which for simplicity we can assume to be $|0\rangle$ (if this is not an appropriate initial state then the cloning device can prepare a more appropriate state starting from $|0\rangle$). Copying the two basis states requires the process

$$|0\rangle|0\rangle \longrightarrow |0\rangle|0\rangle \qquad |1\rangle|0\rangle \longrightarrow |1\rangle|1\rangle \tag{6.5}$$

and this is easily implemented as it can be achieved by a controlled-NOT gate. However this approach *cannot* be used to clone a general state, such as

$$|\psi\rangle = \alpha|0\rangle + \beta|1\rangle. \tag{6.6}$$

If a controlled-NOT gate is used to "copy" this state, the result will be

$$|\psi\rangle|0\rangle = \alpha|0\rangle|0\rangle + \beta|1\rangle|0\rangle \longrightarrow \alpha|0\rangle|0\rangle + \beta|1\rangle|1\rangle, \tag{6.7}$$

which follows immediately from the linearity of the controlled-NOT gate. This state should be compared with the desired state, which has the form

$$|\psi\rangle \otimes |\psi\rangle = \alpha^2|0\rangle|0\rangle + \alpha\beta|0\rangle|1\rangle + \beta\alpha|1\rangle|0\rangle + \beta^2|1\rangle|1\rangle. \tag{6.8}$$

Clearly these states are only the same in the extreme limits of $\alpha = 1$ or $\beta = 1$, that is when the state being copied is a basis state. This result suggests that quantum cloning is indeed impossible, but is not completely convincing as one particular operation for cloning the basis states has been assumed. However, any other putative cloning method is fundamentally equivalent, and the argument proceeds by linearity as before. Just as was the case for characterizing a quantum state, it is possible to optimize the cloning process to work for another particular pair of states, but no solution exists for a general unknown state.

Accepting that an unknown quantum state cannot be copied accurately, one might still ask whether the output of a quantum cloner could in any way assist an attempt to characterize an unknown state. In fact, the form of the state in equation (6.7) immediately rules this out, as this is an entangled state, in which the properties of the two qubits are completely correlated. Once the first qubit has been measured we know that the second qubit will have the same state; actually measuring this state tells us nothing new about the system.

There is in fact an important link between the problem of measurement and the no-cloning theorem. The fact that an unknown quantum state cannot be copied means that the measurement problem cannot be overcome by copying. Similarly, the inability to accurately characterize an unknown state rules out an obvious cloning strategy: measuring the state precisely and crafting identical copies.

6.3 Fidelity

The discussions above showed that it is impossible to perform certain operations (measurement and copying) on quantum bits without the possibility of error. The obvious next question is how accurately these operations can be performed. Critical to this question is the concept of *fidelity*, which measures how close two states (or, by extension, two operations) are to one another.

For assessing state fidelity it seems obvious that the measure should be built around the inner product. Since the two basis states are orthonormal, we know that

$$\langle 0|0 \rangle = \langle 1|1 \rangle = 1 \qquad \langle 0|1 \rangle = \langle 1|0 \rangle = 0 \tag{6.9}$$

which makes sense, as $|0\rangle$ and $|1\rangle$ are exactly identical to themselves, and completely different from each other. However the inner product of two general states will be a complex number, while fidelity should be a real number between 0 and 1. A better definition of the fidelity of one ket $|\phi\rangle$ with respect to another ket $|\psi\rangle$ is

$$F(|\phi\rangle, |\psi\rangle) = |\langle \phi|\psi \rangle|^2 = \langle \psi|\phi \rangle \langle \phi|\psi \rangle. \tag{6.10}$$

This definition can be extended to measure the fidelity between a pure state $|\psi\rangle$ and a mixed state described by a density matrix ρ

$$F(\rho, |\psi\rangle) = \langle \psi|\rho|\psi \rangle, \tag{6.11}$$

which clearly reverts to the original form for two pure states, when $\rho = |\phi\rangle\langle\phi|$. Note that some authors use the square root of this definition, which can also be extended to measure the fidelity between two mixed states.

To take a simple example, consider the fidelity between a general state and itself:

$$F = \begin{pmatrix} \alpha^* & \beta^* \end{pmatrix} \begin{pmatrix} \alpha\alpha^* & \alpha\beta^* \\ \beta\alpha^* & \beta\beta^* \end{pmatrix} \begin{pmatrix} \alpha \\ \beta \end{pmatrix} \tag{6.12}$$

$$= (|\alpha|^2 + |\beta|^2)^2 \tag{6.13}$$

$$= 1. \tag{6.14}$$

A more interesting case is the fidelity between an arbitrary pure state and the same state after a measurement in the computational basis:

$$F = \begin{pmatrix} \alpha^* & \beta^* \end{pmatrix} \begin{pmatrix} \alpha\alpha^* & 0 \\ 0 & \beta\beta^* \end{pmatrix} \begin{pmatrix} \alpha \\ \beta \end{pmatrix} \tag{6.15}$$

$$= |\alpha|^4 + |\beta|^4 \tag{6.16}$$

$$= 1 - 2|\alpha|^2|\beta|^2. \tag{6.17}$$

Clearly the process of measurement damages a state unless the state is a basis state of the measurement (so that either α or β is equal to zero). The worst case occurs for states like $|\pm\rangle$ for which $|\alpha| = |\beta| = 1/\sqrt{2}$, resulting in a fidelity of $F = 1/2$. For the general case it can be shown that the average fidelity of a state after a measurement is $2/3$.

This result can be interpreted in two different ways. Firstly, as noted above, measuring an unknown state will damage it, and so it is always possible to check whether someone has been looking at your "secret" state. (This fact underlies the idea of *quantum money*, which was invented by Steven Wiesner long before quantum information theory was thought of, and ultimately underlies *quantum cryptography*.) Secondly, the state after the measurement also describes the state of knowledge of the person who performed the measurement, and this shows that their knowledge of the state can never be perfect unless the correct measurement basis was used.

6.4 Local operations and classical communication

So far we have only considered two-qubit states where both qubits are accessible to us. The situation becomes much more interesting when different qubits are controlled by different people.

Consider two people, traditionally called Alice and Bob, each of whom have one qubit (that is, they have possession and control of the physical system used to implement the qubit) of a two-qubit system. We assume that they can both manipulate their *own* qubit in any way they desire: they can apply single-qubit logic gates, make measurements, etc., but they have no direct access to the other person's qubit. Speaking technically, we say that Alice and Bob have access to the complete set of *local* operations. We also assume that Alice and Bob can communicate by sending classical messages, reporting the results of measurements on their own qubits, or asking that certain gates be applied to the other person's qubit. This set of abilities is described as *local operations and classical communication*, usually abbreviated to LOCC. By local, here, we mean local in the obvious everyday sense, rather than in the relativistic sense; however, the extension to include classical communications, which we assume to be limited by the speed of light, means that LOCC is equivalent to relativistically local.

If the two-qubit system is in a separable state then nothing mysterious occurs. Recall that it is the nature of a separable state that the two qubits have individual properties, and it makes sense to treat them as individual objects. If the two qubits are entangled, however, then the situation is entirely different! It is no longer really possible to talk about the two qubits as separate objects. As a simple example, suppose Alice and Bob share a pair of qubits in the entangled Bell state

$$|\phi^+\rangle = (|00\rangle + |11\rangle)/\sqrt{2} \tag{6.18}$$

and that Alice applies a NOT gate to her qubit (assumed to be the qubit listed first in our notation). The result is a different Bell state,

$$(|10\rangle + |01\rangle)/\sqrt{2} = |\psi^+\rangle. \tag{6.19}$$

In a similar way, Alice can convert the combined two-qubit state into any one of the four Bell states, and Bob can do the same thing. It is no longer possible to divide up the state into portions controlled by Alice and portions controlled by Bob: they both have equal control over the entire state. This behavior lies at the heart of quantum communication protocols such as *quantum dense coding*, which will be explored later; less positively it also makes certain elementary cryptographic operations impossible in the quantum world, most notably the impossibility of *quantum bit commitment*.

Given this key distinction between separable and entangled states, it is reasonable to ask whether Alice (with or without help from Bob) can turn an initially separable state into an entangled state using only local operations and classical communications. It is a key result in quantum information theory that this is impossible, and more generally the amount of entanglement in a quantum system cannot be increased by LOCC.

Example 6.4 A pure state is said to be separable (and therefore not entangled) if it can be written as a direct product of single-qubit states; a mixed state is said to be separable (and therefore not entangled) if it can be written as a mixture of separable pure states. Now suppose that Alice and Bob start with a pair of qubits in the separable state $|0\rangle|0\rangle$, and that they try to create an entangled state by LOCC. Inspired by the standard network, Alice applies a Hadamard to her qubit and then measures it; if she gets a $|0\rangle$ she does nothing, but if she gets a $|1\rangle$ she tells Bob to apply a NOT gate to his qubit. Find the resulting state, and show that it is not entangled.

Solution

After Alice applies her Hadamard, her qubit will be in the state $(|0\rangle + |1\rangle)/\sqrt{2}$; she then makes her measurement and the state decoheres to the mixed state $(|0\rangle\langle0| + |1\rangle\langle1|)/2$. This is all local to Alice, so Bob's qubit is still in the state $|0\rangle$ and we can describe the whole system by the direct product

$$\tfrac{1}{2}(|0\rangle\langle0| + |1\rangle\langle1|) \otimes |0\rangle\langle0| = \tfrac{1}{2}(|00\rangle\langle00| + |10\rangle\langle10|). \tag{6.20}$$

Finally we consider the effect of Alice talking to Bob: if she got $|0\rangle$ then Bob does nothing and the state remains $|00\rangle\langle00|$; if she got $|1\rangle$ then Bob applies a NOT gate to his qubit, converting $|10\rangle\langle10|$ into $|11\rangle\langle11|$. Thus the final state is $\tfrac{1}{2}(|00\rangle\langle00| + |11\rangle\langle11|)$. This superficially looks like $|\phi^+\rangle\langle\phi^+|$, but writing it out in matrix form

$$\begin{pmatrix} \tfrac{1}{2} & 0 & 0 & 0 \\ 0 & 0 & 0 & 0 \\ 0 & 0 & 0 & 0 \\ 0 & 0 & 0 & \tfrac{1}{2} \end{pmatrix} \tag{6.21}$$

we see that it is not the same. Indeed it is obviously not entangled as the form $\tfrac{1}{2}(|00\rangle\langle00| + |11\rangle\langle11|)$ is a mixture of $|00\rangle$ and $|11\rangle$, that is a mixture of two separable states.

Example 6.5 What is the state fidelity between the state considered above and each of the four Bell states? Is the resulting state a mixture of Bell states?

Solution

State fidelities are easy to obtain by direct multiplication: for $|\phi^+\rangle$ we get

$$\frac{1}{2}\begin{pmatrix} 1 & 0 & 0 & 1 \end{pmatrix} \begin{pmatrix} \tfrac{1}{2} & 0 & 0 & 0 \\ 0 & 0 & 0 & 0 \\ 0 & 0 & 0 & 0 \\ 0 & 0 & 0 & \tfrac{1}{2} \end{pmatrix} \begin{pmatrix} 1 \\ 0 \\ 0 \\ 1 \end{pmatrix} \tag{6.22}$$

which is $\tfrac{1}{2}$, and the same result is found for $|\phi^-\rangle$. The fidelity with $|\psi^\pm\rangle$ is found to be zero. This suggests that we can write the state as $\tfrac{1}{2}(|\phi^+\rangle\langle\phi^+| + |\phi^-\rangle\langle\phi^-|)$, and this is indeed the case. Thus a mixture of entangled states need not be entangled.

If Alice and Bob wish to use an entangled state they must either create one by applying two-qubit gates (which requires the two qubits to be brought into direct contact), or use a state prepared by some third party. For simplicity we can often assume that Alice prepares the entangled state and gives one qubit to Bob. It is a curious fact about many quantum communication protocols that it does not matter where the entangled state comes from: if a malicious person seeks to cheat by providing the wrong state then this fact can easily be detected.

Further reading

Quantum measurement is well described in standard texts, especially Mermin (2007). The original proof of the no-cloning theorem, Wooters and Zurek (1982), is an easy read. Quantum fidelity is well explored in Nielsen and Chuang (2000). The original paper on quantum money is Wiesner (1983), and the basic idea is explored in Schumacher and Westmoreland (2010).

Exercises

6.1   A 180_x rotation will normally act as a NOT gate, converting $|0\rangle$ to $|1\rangle$. If, however, the operation of the gate is divided into n equal periods, each of which is followed by a measurement in the computational basis, then the effect of the gate will be interrupted, and in the limit of very large n the evolution normally generated by the gate will be completely suppressed (the quantum Zeno effect). Calculate the probability that every measurement will project the quantum state onto $|0\rangle$, and find an approximate formula in the limit of large n.

6.2   Suppose Alice and Bob share an entangled pair of qubits in the state $|\psi^-\rangle$. Find local operations that Bob can use to convert this to the other three Bell states.

6.3   It can be shown that any single-qubit gate can be constructed out of a suitable network of Hadamard gates and $T = \sqrt{S} = \sqrt[4]{Z}$ gates. Use this fact to prove that the singlet state $|\psi^-\rangle$ is unaffected by any bilateral unitary operation.

6.4   If two qubits in the Bell state $|\psi^-\rangle$ are measured in the computational basis they will always disagree. Use the result of the previous exercise to show that the same property holds for $|\psi^-\rangle$ if the two qubits are measured in the *same* basis, whatever basis is chosen. Does this work for the other three Bell states?

PART II

QUANTUM COMPUTATION

7   Principles of quantum computing

In Part II of this book we show how computations can be implemented using quantum systems. As we will see, the differences between bits and qubits, briefly outlined in Part I, lead to some important consequences for quantum computing. We begin with a brief introduction to the fundamental principles underlying quantum computing: specific implementations will be considered in later chapters.

7.1   Reversible computing

While quantum computation in its modern form is still a relatively young discipline, researchers have been interested in the relationship between quantum mechanics and computing for a long time. Early workers were not interested in the ideas of *quantum parallelism*, which will be explored in the next section, but rather in the question of whether explicitly quantum mechanical systems could be used to implement classical computations. In addition to its intrinsic interest, there are two technological reasons why this might be considered an important question.

The first reason is a direct consequence of Moore's laws. After the development of integrated circuits, computing technology began its headlong dash down the twin roads of ever-faster and ever-smaller devices. These two phenomena are closely related: as computing devices must communicate within themselves, and as the speed of information transfer cannot exceed the speed of light, faster computers must indeed be smaller. There is, however, a limit to this process, defined by the atomic scale: once the size of individual transistors becomes comparable with that of atoms, the old-fashioned approach of micro-electronics becomes completely untenable.

This limit is now being approached, but physicists were well aware of its existence long before reaching the limit became a real danger. The fundamental problem is that while macroscopic objects, such as transistors, are described by irreversible physical processes, the behavior of atoms and other very small objects is essentially reversible: entropy and related ideas such as the arrow of time are essentially macroscopic concepts arising from the statistical behavior of large collections of microscopic objects. Conventional computing, based on logic gates such as AND and OR, is an apparently irreversible process, and it is not immediately obvious that it can be carried out at the atomic level. To overcome this it is necessary to show that computing can be carried out in an essentially reversible manner.

The second reason is closely related but more subtle. In addition to being made out of stuff, classical computers consume energy which they convert to heat. Our ever-faster computers perform this transformation ever more rapidly, and modern computers suffer from the twin problems of excessive energy consumption (familiar as the short battery life of laptops) and excessive heat output (seen in laptops which are too hot to be used on laps). This problem could in principle be completely overcome if reversible computing is possible, as physically reversible processes do not consume any energy. Strictly reversible thermodynamic processes happen infinitely slowly, and an infinitely slow computer would not be particularly useful, but it is possible to build essentially reversible devices which combine useful speeds with very low power consumptions.

It is well known from traditional classical (irreversible) computation that any desired logic operation can be built from a network of AND and NOT gates, and to prove that reversible computing is possible in principle it suffices to exhibit reversible versions of these gates. (Strictly speaking, it is also necessary to implement a CLONE gate and a SWAP gate; these points are pursued in the exercises.) The NOT gate is easy, as this gate is intrinsically reversible; in reversible computing it is conveniently written as

$$—\oplus—$$

$$(7.1)$$

which should be familiar. The reversible AND gate appears trickier, as the AND operation has two inputs and only one output, which is obviously impossible in a reversible situation. The solution is simply to *preserve* both inputs, allowing the logical process to be reconstructed. The output must then be placed in an *ancilla* bit, but it cannot simply overwrite the initial value of the ancilla, as that would be irreversible. Instead, the new value of the ancilla must be obtained by reversibly combining the output of the gate with the old value, and this is most simply achieved by using bitwise addition modulo 2 (the XOR gate). This reversible AND gate

$$
\begin{aligned}
a &\; \longrightarrow\!\bullet\!\longrightarrow\; a' = a \\
b &\; \longrightarrow\!\bullet\!\longrightarrow\; b' = b \\
c &\; \longrightarrow\!\oplus\!\longrightarrow\; c' = c \oplus (a \text{ AND } b)
\end{aligned}
$$

$$(7.2)$$

is usually called the Toffoli gate, but is also known as the controlled-controlled-NOT gate, as a NOT gate is applied to the target bit c if and only if *both* control bits are set to 1.

In passing, it is worth noting that reversible logic can perform any desired transformation on a set of input bits, but that it does not provide any means to set the bits into the desired initial states. This is hardly surprising, as initialization is a manifestly irreversible process (it requires the final state of a bit to be the same, whatever the initial state was). Thus, while it is possible to perform arbitrary logic in a reversible manner, absolutely reversible computing is not possible. For this reason a "reversible" computation is normally broken down into an irreversible setup process, followed by reversible logic. The final readout (measurement) process may also be taken as being irreversible if desired.

Any desired logic operation can be achieved given only a sufficient supply of Toffoli gates and bits initialized to the desired states, but it is often useful to consider larger logical units. A key example is provided by reversible function evaluation, which for a function with two input bits and one output bit takes the form

$$
\begin{array}{c}
a \quad \boxed{} \quad a' = a \\
b \quad \boxed{f} \quad b' = b \\
c \quad \oplus \quad c' = c \oplus f(a, b)
\end{array}
\tag{7.3}
$$

sometimes called an f-controlled-NOT gate. Clearly, values of the function can be obtained by setting a and b to the desired inputs, and setting $c = 0$. Functions with more than two input bits can be implemented by the obvious extension of this gate, while functions with more than one output bit can be handled by combining each output bit with its own ancilla.

7.2 Quantum parallelism

Clearly, classical reversible computing can be achieved on a quantum computer by setting the initial states of the qubits to eigenstates representing the desired inputs and then performing the desired sequence of reversible logic gates. However, quantum computers are capable of much more than this! Ultimately this comes from the fact that quantum operations are *unitary*, which means they are both *reversible* and *linear*, and that quantum bits can be found in superposition states.

Consider a quantum network to evaluate a function f. If the input register is in an eigenstate $|x\rangle$, we can write this process as

$$
|x\rangle|0\rangle \xrightarrow{U_f} |x\rangle|f(x)\rangle.
\tag{7.4}
$$

If the input register is in a superposition, the result is trivially deduced by linearity. The initial superposition state is a linear combination of inputs, and the resulting state will be a linear combination of outputs:

$$
\sum_{j=1}^{N} \alpha_j |x_j\rangle|0\rangle \xrightarrow{U_f} \sum_{j=1}^{N} \alpha_j |x_j\rangle|f(x_j)\rangle.
\tag{7.5}
$$

Thus the quantum computer has effectively evaluated the function over all N inputs at the same time. This effect, known as quantum parallelism, underlies all quantum algorithms. Note that if the input register comprises n qubits then it can be placed in a superposition of 2^n states, and so the quantum computer can perform 2^n calculations at once. While this is impressive, it is not immediately clear how useful it is, and this point will be considered below.

Before doing this, we return briefly to the question of universality of logic gates. As described above, the Toffoli gate is universal for reversible computing, meaning that any reversible logic operation can be implemented using only Toffoli gates. Other universal gates are known, but like the Toffoli gate they are all *three*-bit gates, and it can be shown that a three-bit gate is essential: universal computing cannot be achieved using only one- and two-bit reversible classical gates, such as NOT and controlled-NOT. This appears to contradict a claim made earlier, that universal quantum computing can be achieved using only single-qubit and two-qubit gates. The solution to this paradox is, of course, that the Toffoli gate can be constructed out of single-qubit and two-qubit gates, but only if some of these gates are not classical. For example, a Toffoli gate can be implemented using two controlled-NOT gates, two controlled-SQUARE-ROOT-OF-NOT gates, and one controlled-INVERSE-SQUARE-ROOT-OF-NOT gate. These last two gates can themselves be built out of controlled-NOT gates and single-qubit gates.

7.3   Getting the answer out

Although quantum parallelism allows a quantum computer to simultaneously evaluate a function over a vast superposition of inputs, it is not obvious that the result can actually be used in any useful way, as it appears as a superposition of possible inputs and outputs. Suppose a quantum computer is prepared in the final state given in equation (7.5), and the values of the two quantum registers are read out. Like any quantum measurement this can only result in one of the eigenstates of the measurement basis, and if the measurement is performed in the computational basis then the result will be

$$|x_j\rangle|f(x_j)\rangle \qquad\qquad (7.6)$$

for some value of j. The value returned will be chosen at random, with probabilities given by $|\alpha_j|^2$ as usual. Note that the values returned for the input and output will always correspond to the same value of j, as the state before the measurement is an *entangled* superposition of inputs and outputs.

As described above, a quantum function evaluation could be simulated by just evaluating the function over one input chosen at random; clearly this would not be very useful! The secret underlying effective quantum computation is that one does not measure the superposition of function values directly; instead, this is first manipulated to produce an interesting result. To be useful this process must combine the different values of $f(x_j)$ in a suitable fashion, so that the final result depends on all of them. However, as the final measurement outcome can only be a single pair of numbers, this result cannot simply reveal the function values directly. *Quantum computation is all about determining small pieces of information which depend on a large number of intermediate results.*

7.4   The DiVincenzo criteria

The previous sections simply assumed that we had access to a general-purpose quantum computer which we could use to implement our algorithms, and completely ignored how this might actually be done. We will consider some possible implementations in detail in later chapters, but before doing so it is useful to consider the problem in general.

Almost any physical system could in principle be considered as a candidate for implementing quantum computing, but to be a serious candidate a proposed system must have certain properties. A traditional list of essential requirements was described by David DiVincenzo, and his criteria provide a useful structure for discussions.

1. A scalable physical system with well-characterized qubits.
2. The ability to initialize the state of the qubits to a simple fiducial state, such as $|0\rangle$.
3. Long relevant decoherence times, much longer than the gate operation time.
4. A universal set of quantum gates.
5. A qubit-specific measurement capability.
6. The ability to interconvert stationary and flying qubits.
7. The ability to faithfully transmit flying qubits between specified locations.

It is important to note that fulfilling these criteria is only necessary for proposals to build large-scale general-purpose quantum computers; small "toy" computers can be built with systems that do not completely do so. Furthermore, the last two criteria are not in fact required for quantum computers themselves, but rather in order to build a "quantum internet." However, these criteria *are* important if quantum devices are to be used to implement complex quantum communication protocols, such as quantum teleportation, which are described in the last part of this book.

At the current time there are no known physical systems which clearly fulfill all seven (or even just the first five) DiVincenzo criteria. However, there are several systems which fulfill these criteria well enough to make simple demonstrations possible, and it is these that we will concentrate on. For quantum communication it is clearly essential that criterion 7 be fulfilled; this is currently only really achievable for photons, which unsurprisingly dominate this field (and Part III of this book). For quantum computation the one critical requirement is criterion 4, a universal set of logic gates, as without this it is impossible to demonstrate any interesting algorithms. Here an early lead was achieved by trapped atoms and ions and by NMR, which were introduced in Part I of this book, and it is these techniques that we will concentrate on here.

Further reading

There is a good introduction to reversible computing in Feynman (1999), and more detailed treatments can be found in journal articles by Bennett (1973, 1982), Fredkin and Toffoli (1982), and Landauer (1982). The core ideas of quantum computing, including explicit constructions of the Toffoli gate from one- and two-qubit gates, are well covered in standard texts such as Nielsen and Chuang (2000), Estève *et al.* (2003), and Stolze and Suter (2008). The DiVincenzo criteria are well described in journals by DiVincenzo (2000) and Bennett and DiVincenzo (2000).

Exercises

7.1   Show that for reversible classical computing CLONE and SWAP gates can be built out of networks of controlled-NOT gates. Can these networks also be used for quantum computing?

7.2   Show how to build NOT and controlled-NOT gates from Toffoli gates. Design a reversible OR gate using only Toffoli gates and NOT gates.

7.3   The Fredkin gate is a three-bit gate which swaps its two target bits if the single control bit is set to 1. Show how a Fredkin gate can be used to implement reversible NOT and AND gates.

7.4   Use your network for a SWAP gate to show how a Fredkin gate can be built from three Toffoli gates. Is it possible to build a Toffoli gate using only Fredkin gates?

7.5   Explain why the network identity

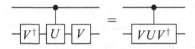

works (note that the apparent reversal in the order of the operators simply reflects the different ordering conventions for operators and networks). Use this identity to construct a Fredkin gate using only a single Toffoli gate and two controlled-NOT gates.

7.6   Consider the reversible half adder network

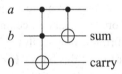

Explain how this network works. Why isn't it necessary to preserve the second input?

8 Elementary quantum algorithms

Although the basic ideas of quantum computation are fairly straightforward, the detailed manipulations required to extract a useful result from an entangled superposition of answers can be rather complicated. There are, however, a small number of algorithms which are simple enough to explain using only elementary methods. The best example is Deutsch's algorithm: as this only requires two qubits its properties are amenable to brute-force matrix calculations, and these calculations can be simplified using a variety of short cuts, which also give some insight into how and why the algorithm works. We will also explore the simplest case of Grover's quantum search algorithm, which is once again simple enough to be tacked by brute force. Finally we will look briefly at two methods used to stabilize quantum computers against errors.

8.1 Deutsch's algorithm

The invention of Deutsch's algorithm can be taken as defining the start of modern quantum computation, and it remains a key example, exhibiting many of the key properties of quantum algorithms in a particularly simple form. (Note, however, that the version of Deutsch's algorithm described here is not in fact the original but a later modification which is both more powerful and easier to understand.)

Consider a binary function f from one bit to one bit, that is a function which takes in either 0 or 1 as its input and returns either 0 or 1 as its output. There are four such functions which may be conveniently labeled by their outputs as shown in Table 8.1. These functions can be divided into two *constant* functions (f_{00} and f_{11}), which have the same output for both inputs, and two *balanced* functions (f_{01} and f_{10}), which have one output of 0 and one of 1. Equivalently, these functions can be classified according to their *parity*, defined as $f(0) \oplus f(1)$.

Deutsch's problem considers the determination of the parity of some unknown function f chosen from these four functions. It is assumed that the only way in which we can access this function is by the use of an *oracle*. This is just a fancy name for a black-box implementation of f which allows us to investigate a function f by asking its value for some input x, but does not permit the internal workings of f to be investigated in any other way. The aim is to find the parity of f with the smallest number of *oracle calls*, that is the smallest number of *queries* about values of $f(x)$.

	Table 8.1	The four binary functions from one bit to one bit		
x	$f_{00}(x)$	$f_{01}(x)$	$f_{10}(x)$	$f_{11}(x)$
0	0	0	1	1
1	0	1	0	1

In the language of quantum computing, this oracle must take the form of a propagator, which performs the transformation

$$|x\rangle|y\rangle \xrightarrow{U_f} |x\rangle|y \oplus f(x)\rangle, \tag{8.1}$$

allowing the function to be evaluated in the usual reversible manner. This can be depicted as an (abstract) quantum circuit

$$\begin{array}{c} |x\rangle \;\text{—}\boxed{}\text{—}\; |x\rangle \\ \quad\quad U_f \\ |y\rangle \;\text{—}\boxed{}\text{—}\; |y \oplus f(x)\rangle \end{array} \tag{8.2}$$

and explicit constructions of the four possible propagators are given below:

$$\tag{8.3}$$

Other constructions are of course possible, but it does not matter *how* a circuit is implemented: all circuits with the same effects are completely equivalent, and so we can choose to analyze these particular implementations. Note that each of these networks can in some sense be considered as an f-controlled-NOT

$$\boxed{U_f} = \begin{array}{c}\boxed{f}\\ \oplus\end{array} \tag{8.4}$$

although $U_{f_{11}}$ is just a simple (uncontrolled) NOT gate, and $U_{f_{00}}$ does not include any gates at all.

If we confine $|x\rangle$ and $|y\rangle$ to the two eigenstates $|0\rangle$ and $|1\rangle$ then we can perform classical reversible computations using this circuit: setting $y = 0$ allows $f(x)$ to be obtained directly, for any desired value of x. Thus we can determine $f(0)$ and $f(1)$, and then combine them to obtain the parity. Classically this is the best we can do: the only way to determine $f(0) \oplus f(1)$ is to find $f(0)$ and $f(1)$ separately, which requires two oracle calls. Using quantum computation, however, we can go beyond this. The parity of the function is only a single bit of information, and a quantum computer can determine this single bit with only a single oracle call using the quantum network given below:

$$\begin{array}{c} |0\rangle \;\text{—}\boxed{H}\text{—}\;\boxed{}\;\text{—}\boxed{H}\text{—}\; |f(0) \oplus f(1)\rangle \\ \quad\quad\quad\quad U_f \\ |1\rangle \;\text{—}\boxed{H}\text{—}\;\boxed{}\;\text{—}\boxed{H}\text{—}\; |1\rangle \end{array} \tag{8.5}$$

This network allows the value of $f(0) \oplus f(1)$ to be read out directly from the *first* qubit, that is the qubit used to define the input in a classical function evaluation. Note, however, that the values of $f(0)$ and $f(1)$ are *not* obtainable. To do this would provide two bits of information, and this requires two oracle calls.

So far we have simply *asserted* that this network will solve Deutsch's problem, and the next step is to show that it does in fact do so. This can be achieved in many different ways, and here we consider a range of possibilities of varying sophistication.

The crudest approach is simply to work out what happens by direct matrix multiplication. To do this we need a matrix description of the two-qubit Hadamard

$$H^{\otimes 2} = H \otimes H = \frac{1}{2}\begin{pmatrix} 1 & 1 & 1 & 1 \\ 1 & -1 & 1 & -1 \\ 1 & 1 & -1 & -1 \\ 1 & -1 & -1 & 1 \end{pmatrix}, \tag{8.6}$$

explicit forms for the four possible propagators

$$U_{f_{00}} = \begin{pmatrix} 1 & 0 & 0 & 0 \\ 0 & 1 & 0 & 0 \\ 0 & 0 & 1 & 0 \\ 0 & 0 & 0 & 1 \end{pmatrix} \quad U_{f_{01}} = \begin{pmatrix} 1 & 0 & 0 & 0 \\ 0 & 1 & 0 & 0 \\ 0 & 0 & 0 & 1 \\ 0 & 0 & 1 & 0 \end{pmatrix}$$

$$U_{f_{10}} = \begin{pmatrix} 0 & 1 & 0 & 0 \\ 1 & 0 & 0 & 0 \\ 0 & 0 & 1 & 0 \\ 0 & 0 & 0 & 1 \end{pmatrix} \quad U_{f_{11}} = \begin{pmatrix} 0 & 1 & 0 & 0 \\ 1 & 0 & 0 & 0 \\ 0 & 0 & 0 & 1 \\ 0 & 0 & 1 & 0 \end{pmatrix} \tag{8.7}$$

and a description of the initial state

$$|01\rangle = \begin{pmatrix} 0 \\ 1 \\ 0 \\ 0 \end{pmatrix}. \tag{8.8}$$

The result then follows by direct multiplication. Note that this can be achieved either by multiplying the ket vector by each matrix in turn, or alternatively by multiplying the three matrices together first, and then applying the resultant matrix product to the ket vector. Both approaches have advantages, and it is not always immediately obvious which is the best approach in any particular case.

8.2 Why it works

Brute-force calculations are sufficient to prove that Deutsch's algorithm does in fact work as stated, but do not give any insight into why it works. A more illuminating approach is to consider the fate of the second qubit. This begins in $|1\rangle$, which is converted to $|-\rangle$ by the first Hadamard, and then evolves according to

$$\frac{|0\rangle - |1\rangle}{\sqrt{2}} \xrightarrow{U_f} \frac{|0 \oplus f(x)\rangle - |1 \oplus f(x)\rangle}{\sqrt{2}} \tag{8.9}$$

where x is the state of the first qubit. Now if $f(x) = 0$ this is just equal to $|-\rangle$, while if $f(x) = 1$ it simplifies to

$$\frac{|1\rangle - |0\rangle}{\sqrt{2}} = -|-\rangle \tag{8.10}$$

and so the process can be summarized as

$$|-\rangle \xrightarrow{U_f} (-1)^{f(x)}|-\rangle. \tag{8.11}$$

The value of $f(x)$ now appears not in the value of the qubit, but rather in its phase. If this phase were a global phase then the information would in effect be lost, but this does not occur in Deutsch's algorithm as the first qubit is also in a superposition state. Thus the algorithm begins with the sequence of transformations

$$|0\rangle|1\rangle \xrightarrow{H^{\otimes 2}} |+\rangle|-\rangle = \frac{|0\rangle|-\rangle + |1\rangle|-\rangle}{\sqrt{2}}$$
$$\xrightarrow{U_f} \frac{(-1)^{f(0)}|0\rangle|-\rangle + (-1)^{f(1)}|1\rangle|-\rangle}{\sqrt{2}}. \tag{8.12}$$

Note that the phase does not "belong" to the second qubit, but to the whole system, and can equally well be thought of as being applied to the first qubit. (This effect, sometimes called *phase kickback*, is very common in quantum computing.)

The second qubit is always in state $|-\rangle$, and this term can be factored out. Now if the function f is constant, so that $f(1) = f(0)$, equation (8.12) simplifies to

$$(-1)^{f(0)} \times \frac{|0\rangle + |1\rangle}{\sqrt{2}} \times |-\rangle = (-1)^{f(0)}|+\rangle|-\rangle \tag{8.13}$$

while if the function is balanced the result is

$$(-1)^{f(0)} \times \frac{|0\rangle - |1\rangle}{\sqrt{2}} \times |-\rangle = (-1)^{f(0)}|-\rangle|-\rangle. \tag{8.14}$$

The initial phase term is a global phase and can be dropped. Finally, applying the last pair of Hadamard gates gives either $|0\rangle|1\rangle$ or $|1\rangle|1\rangle$ as appropriate.[1]

The third approach is to combine the insights obtained from the second approach with knowledge of the properties of gates. A little thought about the propagators U_f reveals that they all take the form of some combination of $\mathbb{1}$ gates (when $f = 0$) and X gates (when $f = 1$) applied to the second qubit, while the first qubit is left untouched. The effect of the Hadamard gates applied to the second qubit before and after U_f is to convert this gate to an equivalent combination of $\mathbb{1}$ and Z gates (since $H\mathbb{1}H = \mathbb{1}$ and $HXH = Z$). Finally, applying

---

[1] The results above suggest that the value of $f(0)$ can be obtained from the *phase* of the final state; since we know the parity this would allow both $f(0)$ and $f(1)$ to be determined in a single step. However, this phase is a global phase, and thus not physically detectable. The quantum algorithm merely provides an efficient way of answering a question; it does not perform the completely impossible.

a Z gate to a qubit in state $|1\rangle$ is equivalent to multiplying the state by minus one. Thus we can immediately deduce that the first qubit undergoes the sequence of transformations

$$|0\rangle \xrightarrow{\text{H}} \frac{|0\rangle + |1\rangle}{\sqrt{2}} \xrightarrow{U_f} \frac{(-1)^{f(0)}|0\rangle + (-1)^{f(1)}|1\rangle}{\sqrt{2}} \xrightarrow{\text{H}} |f(0) \oplus f(1)\rangle \qquad (8.15)$$

while the second qubit is converted from $|1\rangle$ to $|-\rangle$, and back to $|1\rangle$. Note that we do not need to write out the state of the second qubit explicitly at each stage: all the states involved are *separable* and so it makes sense to talk about the state of the two qubits individually.

8.3 Circuit identities

The final, and perhaps the boldest approach is to use operator identities to analyze the whole circuit and only then consider the action on particular qubits. This can be done using explicit propagator identities, or simply by using circuit identities to manipulate the circuits themselves. Thus, for example, the combined propagator for the case of f_{11} is given by

$$(\text{H} \otimes \text{H}).(\mathbb{1} \otimes \text{X}).(\text{H} \otimes \text{H}) = (\text{H}.\mathbb{1}.\text{H}) \otimes (\text{H}.\text{X}.\text{H}) \qquad (8.16)$$

which simplifies to $\mathbb{1} \otimes \text{Z}$ using standard identities. In the same way the circuit for the case of f_{01} can be manipulated using circuit identities

$$(8.17)$$

and so the effect of applying Hadamard gates to both qubits before and after a controlled-NOT gate is simply to reverse the roles of control and target qubits. This greatly reduces the amount of effort, and examining the final circuits for the four possible functions

$$(8.18)$$

allows the final results, including the global phases, to be calculated immediately.

In the case of Deutsch's algorithm it is not immediately obvious that this approach provides much insight into *why* it works, beyond explaining why it is essential that the second qubit begins in state $|1\rangle$. For other cases, however, most notably the Bernstein–Vazirani algorithm discussed in the next chapter, circuit identities seem to provide a great deal of insight into how they work. The essential feature in these cases is once again the ability of Hadamard gates to interconvert control and target qubits. This is not, of course, possible with classical computers as Hadamard gates are not permitted gates for classical bits.

8.4　Deutsch's algorithm and interferometry

There is another interesting way of looking at Deutsch's algorithm, which emphasizes the physics rather than the mathematics. In essence there is an extremely close link between Deutsch's algorithm and a Mach–Zehnder interferometer.

Consider a single photon which is incident on a beam splitter. As usual we treat the beam splitter as a Hadamard gate, taking the photon from state $|0\rangle$ to state $H|0\rangle = |+\rangle$. The two photon paths are then recombined at another beam splitter (Hadamard gate), and the final result will be $H|+\rangle = |0\rangle$. Thus the photon will always emerge at the same port of the second beam splitter. In effect we have reduced the traditional complex treatment of an interferometer to the trivial observation that $HH = \mathbb{1}$.

Now suppose that we introduce phase shifters into the separated beam paths, which apply a phase shift of π, and so multiply a state by -1. If we introduce a phase shifter into the $|1\rangle$ path, then our state $|+\rangle$ will clearly be converted to $|-\rangle$, and the final output will now be $|1\rangle$. If we introduce a phase shifter into the $|0\rangle$ path then the intermediate state will be converted into $-|-\rangle$, and once again the output at the second beam splitter will be $|1\rangle$. Finally, inserting phase shifters into both paths converts $|+\rangle$ to $-|+\rangle$, leading to an output of $|0\rangle$ (neglecting the global phase shift as usual). Thus the output will be $|1\rangle$ if the phase shift on the two paths is *different*, and $|0\rangle$ if the phase shift on the two paths is the same. The analogy with Deutsch's algorithm is obvious.

8.5　Grover's algorithm

Grover's quantum search algorithm is an example of a second great class of quantum algorithms, which are fundamentally different from Deutsch's algorithm and its generalizations explored in the next chapter. The algorithm can be described in many ways, but the simplest approach is once again to consider the analysis of binary functions.

Suppose we have a binary function f with n input bits and a single output bit, and we are guaranteed that $f(x) = 1$ for exactly one input, with $f(x) = 0$ for the remaining $2^n - 1$ inputs (useful but apparently arbitrary guarantees of this kind are usually called *promises*). Beyond this we know nothing about f, and can only obtain more information by making oracle queries. It is conventional to label these functions by stating the value of the unique *satisfying* input for which the function is equal to 1, so, for example, $f_{01}(01) = 1$ while $f_{01}(00) = f_{01}(10) = f_{01}(11) = 0$. Despite the identical names, it is vital to note that these four functions f_{ij} are *not* the same as the four functions f_{ij} used in describing Deutsch's algorithm.

The problem is to find the unique value of x for which $f(x) = 1$ with the smallest number of queries. This mathematical task is equivalent to the problem of finding out someone's name given only their telephone number and a copy of the relevant telephone directory. With a classical computer the only possible approach is to try different inputs (names in the directory) at random until we find the unique satisfying input (telephone number). If the number of possible inputs $N = 2^n$ is small, then this process will be easy, but as n grows the process becomes extremely inefficient: on average it will be necessary to try around $N/2$ inputs, which for $n = 32$ means billions of queries.

Grover's quantum search algorithm allows the satisfying input to be located much more rapidly, with about $\sqrt{N}$ queries. The general case is beyond the scope of this chapter, but the simple case of $n = 2$ is relatively easy to analyze. In this case a classical search will require either 1, 2 or 3 queries (it is never necessary to try all four inputs, as if the satisfying input has not been located in the first three attempts we know that the last input must be the one we are seeking), while Grover's algorithm is guaranteed to locate the satisfying input in a single query by using the quantum network shown below:

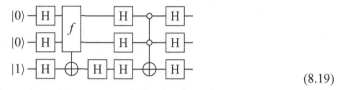

$$(8.19)$$

The first three-qubit gate is an f-controlled-NOT gate. The second one is similar to a Toffoli gate but the empty circles on the control lines indicate that it applies a NOT gate to the target bit if and only if both control bits are in state 0. (As usual these three-qubit gates can be built out of one- and two-qubit gates, but it is simpler to consider them at this more abstract level.)

To simplify the analysis of this network, we begin with the last qubit which is an ancilla. This behaves in very much the same way as the second qubit in the Deutsch algorithm, converting the f-controlled-X gates into f-controlled-Z gates, and thus into phase shifts. As a result we can now ignore this qubit and concentrate our attention on the first two.

The first two qubits begin in the state $|00\rangle$, which is converted by the simultaneous Hadamard gates into a uniform superposition of the four possible inputs

$$|00\rangle \xrightarrow{\mathrm{H}^{\otimes 2}} |+\rangle|+\rangle = \frac{|00\rangle + |01\rangle + |10\rangle + |11\rangle}{2}. \tag{8.20}$$

The f-controlled-Z gate now evaluates the function simultaneously over all the possible inputs and applies appropriate phase shifts to give the state

$$\frac{(-1)^{f(00)}|00\rangle + (-1)^{f(01)}|01\rangle + (-1)^{f(10)}|10\rangle + (-1)^{f(11)}|11\rangle}{2} \tag{8.21}$$

where, because of the promise about f, only one of the states will have a minus sign. To take a concrete example, if the satisfying input is 01, then the state will be

$$\frac{|00\rangle - |01\rangle + |10\rangle + |11\rangle}{2}. \tag{8.22}$$

The desired satisfying state has now been picked out, and it might seem that the problem has been solved, but this is not in fact the case. Although the satisfying state is uniquely labeled by its phase, it still contributes the same absolute amplitude to the superposition as the other states, and so any attempt to measure the state of the first two qubits will simply return one of the possible inputs at random.

The purpose of the remaining gates is to convert this phase difference into an amplitude difference. This is most easily proved by direct matrix multiplication, an approach explored in the exercises, but for the moment it suffices to note that in the network above these gates will concentrate *all* the amplitude in the superposition on the desired state, so that a measurement of the first two qubits will immediately reveal the satisfying input. If $n > 2$ this process cannot be achieved in a single step, and it is necessary to use a sequence of oracle queries and *amplitude amplification* steps, but the quantum search is still much more efficient than its classical equivalent. To make further progress requires a more advanced treatment, which is explored in the next chapter. Before this, however, we will look briefly at two methods for stabilizing quantum computers against errors.

8.6   Error correction

The discovery of quantum error correction was one of the key developments in quantum computing, as it convinced many skeptics that quantum computing might just be possible. All computers are vulnerable to errors, but this is much more true of quantum than classical devices, as classical digital computers are inherently stabilized. Classical bits can only take the values 0 and 1, and if the physical implementation of the bit (such as a voltage) wanders away from its ideal value it can be driven back. This is impossible for quantum computers for two reasons. Firstly, quantum bits are not confined to $|0\rangle$ and $|1\rangle$, but will be found in general superposition states. Secondly, the processes that drive a bit back to its ideal state are dissipative, and so non-unitary, while quantum evolution must be unitary.

An alternative approach to handling errors is to use error correcting codes. For example, it is possible to encode a single *logical bit* by using *code words* made up of three *physical bits*, with 000 representing the logical bit 0, and 111 representing logical 1. If any one physical bit is corrupted this can be detected, as the three bits will no longer have the same value, and setting the single bit that disagrees back to the consensus state will correct the error. A more careful analysis shows that as long as bits become corrupted independently and with a small error probability this can provide effective error suppression, and more complex schemes can give even better performance.

This classical scheme is useless in the quantum world, as it relies on repeatedly measuring the values of the physical qubits and comparing them. For qubits this will destroy the fragile quantum information stored in superposition states. The key insight for quantum error correction, however, is that it is in fact possible to perform these measurements in such a way that the error is identified without damaging the superposition. To achieve this it is essential that the measurement *only* provides information about the error; it must provide no information at all about the state of the logical qubit.

As a concrete example we consider a system where one logical qubit is encoded in three physical qubits, where one of the qubits may have been damaged by a *spin flip error*, that is one of the three qubits may have experienced an unintended NOT gate (X gate). As for the classical code, the two code words used are $|0_L\rangle = |000\rangle$ and $|1_L\rangle = |111\rangle$, and an arbitrary superposition state is encoded as

$$|\psi_L\rangle = \alpha|0_L\rangle + \beta|1_L\rangle = \alpha|000\rangle + \beta|111\rangle \tag{8.23}$$

which can be achieved using either of the encoding networks shown below:

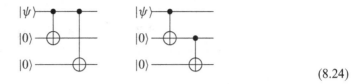

$$\tag{8.24}$$

After the error process (we assume that at most one of the three physical qubits has been flipped), this state is converted to

$$|\psi_L\rangle \longrightarrow \begin{cases} \alpha|000\rangle + \beta|111\rangle & \text{no error} \\ \alpha|100\rangle + \beta|011\rangle & \text{bit 1 flipped} \\ \alpha|010\rangle + \beta|101\rangle & \text{bit 2 flipped} \\ \alpha|001\rangle + \beta|110\rangle & \text{bit 3 flipped} \end{cases} \tag{8.25}$$

depending on the exact form of the error. The task is to identify the error while learning nothing about α and β. This can be achieved using the following network, which requires two additional (ancilla) qubits:

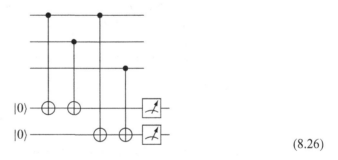

$$\tag{8.26}$$

where the bottom two qubits are the ancillas. Note that only the ancillas are directly measured, not the logical qubit! If no error has occurred then both ancillas will end up in state $|0\rangle$, while if an error has occurred then either or both of the ancillas will end up in state $|1\rangle$. The first ancilla can only end up in state $|1\rangle$ if the physical qubits 1 and 2 have different values, while the second ancilla can only end up in state $|1\rangle$ if the physical qubits 1 and 3 have different values. By this means the error can be detected and corrected; a more complete analysis is left to the exercises. Finally, the corrected code word can be converted back to a single qubit by reversing the encoding circuit.

A very similar approach can be used to correct for random *phase flip* errors, that is random Z gates, by using the code words $|0_L\rangle = |{+}{+}{+}\rangle$ and $|1_L\rangle = |{-}{-}{-}\rangle$. It is also possible to correct for both sorts of error at the same time. The conceptually simplest approach is to use a nine-qubit code which concatenates the two types of error correction described above; more efficient methods are also available, but these are much more complex to explain.

So far we have only considered gross errors, taking the form of X gates and Z gates, and have not considered more subtle errors, such as small rotations around arbitrary axes. Remarkably, the methods outlined above are sufficient to correct any such error. This point is briefly explored in the exercises.

8.7 Decoherence-free subspaces

The key assumption about errors used in quantum error correction is that they are *independent* and *uncorrelated*, that is the probability of any one qubit being affected does not depend on what has happened to other qubits. In practice, however, errors are caused by undesirable interactions with the environment, and in many implementations it is likely that the errors will be highly correlated: two qubits which are physically close in space will have similar environments and thus similar unwanted interactions. This is a problem for quantum error correction, but can be exploited to give an entirely different method of tackling errors, based on the idea of a decoherence-free subspace or DFS.

This method has the advantage that its implementation does not require projective measurements, and so it can be used with techniques where this is difficult, such as NMR. Furthermore, error correction requires that errors be constantly detected and corrected, with gates being applied to every qubit in the system on a regular basis. For large systems this means that gates *must* be applied in parallel to multiple qubits at the same time. By contrast, the DFS approach aims to prevent errors from happening in the first place, and is intrinsically parallel.

Once again the method relies on the use of code words. Here we give a very simple description of a subspace resistant to phase flip errors, motivated by implementations in NMR. Consider the two code words

$$|0_L\rangle = (|01\rangle + |10\rangle)/\sqrt{2} \qquad |1_L\rangle = (|01\rangle - |10\rangle)/\sqrt{2} \tag{8.27}$$

which are the $|\psi^\pm\rangle$ Bell states. These are orthogonal quantum states, and so can be used to encode a logical qubit. They have the important property that

$$\left(\frac{|01\rangle \pm |10\rangle}{\sqrt{2}}\right) \xrightarrow{Z \otimes Z} -\left(\frac{|01\rangle \pm |10\rangle}{\sqrt{2}}\right) \tag{8.28}$$

so that these states are invariant (up to a global phase) under *simultaneous* Z rotations. As this global phase is the same for both $|0_L\rangle$ and $|1_L\rangle$, a superposition of $|0_L\rangle$ and $|1_L\rangle$ will also have this property, and such a qubit will be invulnerable to perfectly correlated phase decoherence. Invulnerability to correlated X rotations can be achieved using the two Bell states $|\phi^+\rangle$ and $|\psi^+\rangle$ as the logical basis, and complete invulnerability can be achieved using more complex codes.

Note that the encodings given above are not the only possible encodings for decoherence-free subspaces. The simplest way to see this is to note that as general superpositions of $|0_L\rangle$ and $|1_L\rangle$ are decoherence free, then we can choose *any* orthonormal pair of superposition states as our basis states. For simultaneous Z rotations a particularly simple encoding is $|0_L\rangle = |01\rangle$ and $|1_L\rangle = |10\rangle$.

From the description above the decoherence-free subspace approach looks like a magic bullet, but this is an overoptimistic view. The DFS approach only works if errors are perfectly correlated, and this is unlikely to be completely true in practice. In reality a DFS will only resist the correlated part of the errors, and if there are any significant uncorrelated errors these will soon build up. For this reason many researchers believe that the DFS approach should be combined with standard error correction.

A second more subtle point is that it is necessary to implement quantum logic gates on the *logical* qubits, rather than the physical qubits, which requires that much more complicated gates be designed. An alternative approach is to leave the DFS temporarily, while the gate is implemented, and return to it to allow the qubit to be stored. Both methods have been explored.

Further reading

Quantum algorithms are covered in most standard texts, such as Nielsen and Chuang (2000), Estève *et al.* (2003), and Stolze and Suter (2008), but Mermin (2007) provides a particularly good introduction to the circuit identity approach for analyzing algorithms. The original Deutsch algorithm (Deutsch, 1985) is never used, being effectively superseded by a more effective variant (Cleve *et al.*, 1998); by contrast, Grover's quantum search algorithm remains quite close to its original form (Grover, 1997).

The principles of error correction and decoherence-free subspaces are best learned from textbooks, but more advanced concepts, such as fault-tolerant computation, are well described in journals (Knill *et al.*, 1998).

Exercises

8.1 Given an oracle U_f implementing a function from one bit to one bit, design a classical circuit to directly determine the parity of f in two queries using only two bits (that is you may not store results "offline" for later comparisons and you may not use additional ancilla bits).

8.2 Equation (8.3) gives explicit circuits corresponding to the four functions f_{ij} in Deutsch's algorithm. Find an alternative circuit for f_{10} which applies single-qubit gates to the upper qubit rather than the lower qubit.

8.3 Prove the remaining circuit identities shown in equation (8.18).

8.4 Calculate an explicit matrix form for the Grover amplitude amplification operator in the case $n = 2$, and hence show that Grover's quantum search will reveal a single satisfying input in a single query (you may neglect the ancilla qubit and use an explicit phase shift form for the action of the controlled gate on the two main qubits).

8.5   What happens in Grover's algorithm if the function has *two* satisfying inputs? What about three?

8.6   Show that the two encoding networks for quantum error correction given in equation (8.24) will act as desired, and write down corresponding decoding networks.

8.7   Consider the three-qubit spin flip error correcting network shown in equation (8.26). By working through the network, find kets describing the state of the device immediately before the ancilla qubits are measured for an arbitrary logical input with each of the three single-qubit errors or no error. Show that these states can be written as product states of the logical qubit and the ancilla qubits, and hence that measuring the ancillas has no effect on the logical qubit.

8.8   Give explicit forms for the error correcting steps in the three-qubit spin flip error correcting network (that is, what correction operators should be applied for each correction outcome). Show how this process can be replaced by quantum control (replacing measurements and optional gates by conditional logic gates); state two disadvantages of this latter approach.

8.9   What happens to a classical bit protected with a three-bit code if two-bit flip errors occur? What happens in the quantum case?

8.10   Consider a single qubit which starts in the pure state $|\psi\rangle = \alpha|0\rangle + \beta|1\rangle$ and then undergoes one of two sorts of decoherence: (a) rotation around the z axis through an angle of either ϕ or $-\phi$, chosen at random; (b) experiencing a Z gate with probability p or being left alone with probability $1 - p$. Show that the density matrix description of these two cases is fundamentally the same, and determine the relationship between p and ϕ.

9 More advanced quantum algorithms

Deutsch's algorithm and Grover's quantum search are theoretically interesting, but it is not obvious that these two algorithms are actually useful for anything important. We next consider a selection of more advanced algorithms, several of which may have real-life applications. Some of these will be too complicated to explain fully, and their properties will only be sketched briefly.

9.1 The Deutsch–Jozsa algorithm

Deutsch's algorithm is simple, but important, as it shows that a quantum algorithm can find a property of an unknown function (its parity) with a smaller number of queries than any possible classical algorithm (one rather than two). For this reason we can say that quantum computing is more efficient than classical computing within the oracle model of function evaluation. (It is widely believed that quantum computing is more efficient than classical computing in general, but this is a surprisingly hard thing to prove.) The simplicity of the algorithm is also an advantage, as it can be implemented on very primitive quantum computers. Beyond this, however, Deutsch's algorithm is also important as the simplest member of a large family of quantum algorithms, including most notably Shor's quantum factoring algorithm.

The second simplest algorithm in the family is the Deutsch–Jozsa algorithm, which solves a very closely related problem. Consider an unknown binary function with n input bits, giving $N = 2^n$ possible inputs, and a single output bit. For the case $n = 1$, which we analyzed above, such functions are always constant or balanced, but for $n > 1$ this need not be true: for example, a function with $N = 4$ might return the value 0 for one of its inputs and the value 1 for the other three. Suppose, however, that the function is promised to be *either* constant or balanced. Then the Deutsch–Jozsa problem is to determine whether the function is constant or balanced with the smallest number of queries (oracle calls).

The best possible classical algorithm in this case is to simply try inputs at random and compare the outputs. If any two different inputs give different outputs then we can immediately deduce that the function is *balanced*, but if all the outputs are the same it seems likely that the function is *constant*. In this latter case we cannot be sure the function is constant until $N/2 + 1$ different inputs have been tried, by which time a balanced function is *certain* to have revealed itself. Thus solving the Deutsch–Jozsa problem classically will require between 2 (best case) and $N/2 + 1$ (worst case) queries; for the Deutsch problem

$N = 2$ and these two limits are the same. Remarkably, a quantum computer implementing the Deutsch–Jozsa algorithm can *always* answer the question in a single query.

The quantum network for the Deutsch–Jozsa algorithm in the case $n = 2$ is as shown below:

$$|0\rangle -\boxed{H}\ \boxed{}\ \boxed{H}\ \measuredangle$$
$$|0\rangle -\boxed{H}\ \boxed{f}\ \boxed{H}\ \measuredangle$$
$$|1\rangle -\boxed{H}\ \oplus\ \boxed{H}\ -\ |1\rangle$$

(9.1)

where the last qubit is an ancilla, and the result will be read out from the upper two "input" qubits. The similarity with the network for Deutsch's algorithm, and the generalization to cases with $n > 2$ should be obvious, and the analysis is quite similar.

As before the effect of the ancilla bit is simply to convert the f-controlled-NOT to an f-controlled-Z, which is evaluated over a uniform superposition of all possible inputs. Thus the state of the two input qubits immediately before the final Hadamard gates is

$$\frac{(-1)^{f(00)}|00\rangle + (-1)^{f(01)}|01\rangle + (-1)^{f(10)}|10\rangle + (-1)^{f(11)}|11\rangle}{2} = \frac{1}{2}\sum_{x=0}^{3}(-1)^{f(x)}|x\rangle,$$

(9.2)

where if the function is constant all the components in the superposition have the same sign, which can be either plus or minus, and if the function is balanced half the components have a plus sign and the other half a minus sign. Note that up to this point the behavior of the algorithm is identical to Grover's quantum search, except that the distribution of plus and minus signs is different because of the different promise made about the nature of the function f.

The effect of the final set of Hadamard gates is to combine the different amplitudes of these components, allowing the desired answer to be extracted. A full treatment requires a more sophisticated view of the Hadamard gate, which will be given later, but it is obvious from equation (8.6) that the amplitude of the $|00\rangle$ component is simply half the sum of all the amplitudes. If the function is constant then this amplitude will be either $+1$ or -1, and so all the amplitude in the superposition will be concentrated onto $|00\rangle$. In this case a measurement of the input qubits will *certainly* give the result $|00\rangle$. By contrast, if the function is balanced then the amplitude of the $|00\rangle$ component will be zero, and a measurement of the input qubits *cannot* give this result.

By this means balanced functions can be reliably distinguished from constant functions, and this property will generalize to cases with $n > 2$. If the function is neither constant nor balanced then the amplitude of $|00\rangle$ will be non-zero but will not equal ± 1; such functions can appear to be either constant or balanced and so cannot be reliably distinguished.

9.2 The Bernstein–Vazirani algorithm

The Bernstein–Vazirani algorithm once again considers the analysis of functions with n input bits and a single output bit, but the promise about the nature of the function is different. In this case it is promised that

$$f(x) = a \cdot x, \tag{9.3}$$

where a is an n-bit integer labeling the particular function and $a \cdot x$ is a bitwise dot product, calculated by multiplying corresponding bits of a and x and then adding the resulting bits modulo 2. The task is to determine the value of a using the smallest number of queries to evaluate particular values of $f(x)$. Although there are 2^n possible values of a, it can be determined on a classical computer in only n queries. The simplest approach is to choose values of x corresponding to powers of 2, so that their binary representations are zero everywhere except at one bit; in this case the value of $f(x)$ will immediately reveal the corresponding bit of a.

The Bernstein–Vazirani algorithm, however, allows a quantum computer to determine the value of a in a single function evaluation. As usual we assume that the function is evaluated using a quantum oracle which implements the function as an f-controlled-NOT; the quantum network is then *identical* to the network for the Deutsch–Jozsa algorithm, equation (9.1), and the result (the value of a) can be read out from the input qubits as before. The analysis of why this works, however, is quite different, and relies on the fact that *any* function of the form of equation (9.3) can be implemented using only controlled-NOT gates, controlled by the individual input qubits and targeting the ancilla qubit. Explicit propagators for the case $n = 2$ are shown below:

$$\tag{9.4}$$

where the propagator can include either of the two bracketed sections, both of them, or neither. In particular, a controlled-NOT gate controlled by a given input qubit is applied if and only if the corresponding bit in a has the value 1.

The effect of the initial and final Hadamard gates is simply to reverse all these controlled-NOT gates

$$\tag{9.5}$$

using the circuit identity established in equation (8.17). As the ancilla qubit starts in $|1\rangle$ these controlled-NOT gates can simply be replaced by conventional NOT gates, and so the effect of the whole network simplifies to applying NOT gates to the input qubits if and only if the corresponding bit of a is equal to one.

9.3 Deutsch–Jozsa and period finding

The Bernstein–Vazirani algorithm is quite impressive, but only works for a very restricted group of functions, of the form of equation (9.3). The Deutsch–Jozsa algorithm is, perhaps, more general, although in its simplest form it only reveals whether a function is constant or balanced. A more detailed analysis, however, indicates that, for balanced functions, the output provides more information about the pattern of satisfying answers.

The Deutsch–Jozsa algorithm gets its power not just from the fact that quantum parallelism is used to evaluate the function for all its possible inputs in one step, but also from the final Hadamard gates, which act to combine these results in a useful way. We have already seen how the final amplitude of the $|0\ldots00\rangle$ state depends on the sum of all the individual amplitudes, and so the probability of finding all the input qubits in state $|0\rangle$ after the final measurement will be one for constant functions and zero for balanced functions. For balanced functions the amplitude will be distributed over the remaining possible states of the input qubits, and the exact result of the final measurement will depend on how this redistribution occurs.

Example 9.1 For the case $n = 2$ it is relatively simple to see how this occurs by examining the individual rows of $H^{\otimes 2}$, which takes the general form [see equation (8.6)]

$$\begin{pmatrix} + & + & + & + \\ + & - & + & - \\ + & + & - & - \\ + & - & - & + \end{pmatrix}. \tag{9.6}$$

The amplitude can only end up in state $|00\rangle$ if all the phases in the initial superposition are the same (whether they are all positive or all negative). Similarly, the amplitude will only end up in state $|01\rangle$ if the phase *alternates* in the initial superposition, matching the pattern of plus and minus signs in the second row of the matrix, which in turn requires that alternate values of f must equal 0 and 1. For balanced functions, with two adjacent values of f being the same, the amplitude will end up in state $|10\rangle$ or $|11\rangle$, depending on exactly how these values are arranged.

For functions with $n > 2$ the situation is conceptually similar, but in this case some balanced functions will result in amplitudes being distributed over several states in the final superposition, but the key result, that no amplitude can end up on state $|0\ldots00\rangle$, remains true.

The action of the final Hadamard gates can be thought of as extracting the frequency of the phase variation in the initial superposition, and in particular distinguishing between zero frequency for constant functions and some non-zero frequency, or combination of frequencies, for balanced functions. Fundamentally this is performed using a Hadamard transform, sometimes called a Walsh–Hadamard transform. This is a close relative of the Fourier transform, but rather than decomposing a function in terms of smooth sine and cosine functions it uses Walsh functions, sometimes called sal and cal, which are rectangular waveforms, only taking the values $+1$ and -1. The general action of the n-bit Hadamard transform is given by

$$|x\rangle \xrightarrow{\text{H}^{\otimes n}} \frac{1}{2^{n/2}} \sum_{y=0}^{2^n-1} (-1)^{x \cdot y} |y\rangle, \qquad (9.7)$$

where the dot indicates a bitwise dot product as before.

Example 9.2 Various identities can be used to simplify Hadamard transform calculations. In particular

$$(-1)^{(x \oplus a) \cdot y} = (-1)^{(x \cdot y) \oplus (a \cdot y)} = (-1)^{x \cdot y} \times (-1)^{a \cdot y} \qquad (9.8)$$

and so on.

A more powerful example of the Hadamard transform is provided by Simon's period-finding algorithm, which considers the analysis of binary functions from n bits to m bits, which are periodic in the sense that all the values of $f(x)$ are different except that

$$f(x) = f(x \oplus a) \qquad (9.9)$$

for all values of x; in other words, $f(x) = f(y)$ if and only if $y = x \oplus a$. The problem is, of course, to determine the period a. This is straightforward if the function takes some simple form, like a sine wave, but much more complicated if the function has no obvious pattern other than its periodicity.

The best classical method for finding this period is simply to evaluate $f(x)$ for various inputs x and then compare these results. Since $f(x)$ is promised to be different for all values of x except those separated by the period a it suffices to find a single pair of repeated values to locate the period. Assuming that the period is large this will require many different function evaluations: the full calculation is complex, but typically around $2^{n/2}$ evaluations will be required. Clearly this number increases very rapidly with n, and finding the period is a very challenging problem.

By contrast, Simon's quantum period-finding algorithm uses the properties of the Hadamard transform to extract this period much more quickly. The full details of the algorithm are quite complex, but in essence it produces a quantum state of the form

$$\frac{|x\rangle + |x \oplus a\rangle}{\sqrt{2}} \qquad (9.10)$$

for some randomly chosen number x. Clearly this superposition encodes the desired value, but it cannot be obtained directly. If the quantum state is simply measured then either

x or $x \oplus a$ will be returned at random, and neither of these values allows a to be determined unless the other value is also known. The obvious approach, repeating the algorithm, is pointless as a different random value of x will be found in each case. The solution is to apply a Hadamard transformation, giving

$$\frac{1}{2^{(n+1)/2}} \sum_y \left[(-1)^{x \cdot y} + (-1)^{(x \oplus a) \cdot y} \right] |y\rangle = \frac{1}{2^{(n+1)/2}} \sum_y (-1)^{x \cdot y} \left[1 + (-1)^{a \cdot y} \right] |y\rangle, \quad (9.11)$$

and the amplitude of each term in the sum will be zero unless $a \cdot y = 0$. Measuring the quantum state will, therefore, return a randomly chosen number y with the property that $a \cdot y = 0$, acting to narrow down the set of possible values for a. If this process is repeated a new random value of y will be obtained, providing further information about a, and after around n repetitions of the algorithm the value of a will be fully determined.

Example 9.3 Consider the case $n = 4$, and suppose that the first random result is $y = 5$, which is 0101 written as a four-bit binary number. In this case we can deduce that bits zero and two of a must be the same, ruling out half the possible values of a. The second repetition will provide further information about a, unless it happens to return $y = 5$ again (in which case we learn nothing new) or it happens to return $y = 0$ (in which case we learn nothing at all); in all other cases it will again rule out half the remaining possibilities. By contrast, with the classical calculation the first function evaluation reveals nothing about a, and the second evaluation only rules out a single value of a (except in the very unlikely case that the second evaluation happens to locate the period immediately).

9.4 Fourier transforms and quantum factoring

Finally we turn to Shor's quantum factoring algorithm, which triggered a huge expansion in interest in quantum computation. This algorithm has the potential to destroy many current forms of cryptography known as public key methods. These schemes use a public key, which is in essence the product of two very large prime numbers, and a private key, which is derived from the two numbers themselves.[1] The security of the system is based on the apparent difficulty of deducing the private key from the public key, and so is ultimately based on the belief that factoring large composite numbers is very difficult. All currently known factoring schemes on classical computers are inefficient, and it is widely believed that no efficient classical algorithm exists. By contrast, Shor's quantum algorithm is known to be efficient, and so factoring will become easy if a quantum computer is built.

Shor's factoring algorithm is quite complicated, but the basic idea is fairly easy to understand. Given an unknown composite number N it does not seek to factor N directly,

---

[1] The scheme described here, known as RSA, is not the only public key system, but the great majority of systems in widespread use are closely related to it, and can be broken either by Shor's algorithm or minor variations. There are, however, a small number of public key algorithms which have not yet been broken by quantum computers.

but solves a closely related problem based on *modular exponentiation*. Begin by choosing a random number a which is *coprime* with N, that is a number which shares no common factors with N. This check is easily made using Euclid's efficient algorithm for calculating the greatest common divisor (gcd) of a and N, also known as the highest common factor, as $\gcd(a, N) = 1$ if the numbers are coprime. Now consider the function

$$a^x \bmod N \tag{9.12}$$

which has a period r, so that r is the smallest integer such that

$$a^r \bmod N = a^0 \bmod N = 1. \tag{9.13}$$

Finding this period is exactly the sort of thing quantum computers are good at! As described previously, a quantum computer can evaluate the modular exponentiation function for all possible values of x and then attempt to pick out the period r. Rather than using Hadamard transforms, Shor's algorithm uses a more powerful alternative, the n-bit quantum Fourier transform defined by

$$|x\rangle \xrightarrow{\text{QFT}_n} \frac{1}{2^{n/2}} \sum_{y=0}^{2^n-1} e^{i2\pi xy/2^n} |y\rangle, \tag{9.14}$$

where xy is calculated using ordinary integer multiplication. Just like a conventional discrete Fourier transform, this concentrates most of the amplitude on values of y close to the desired period r, and measuring the final state will return r with high probability. Although the quantum Fourier transform looks complicated, it can be implemented using a surprisingly simple network of quantum gates.

The final stage of the algorithm relies on further results from classical mathematics called the *Chinese remainder theorem* and *Fermat's little theorem*. Briefly these can be used to show that if r is even and if $a^{r/2} \bmod N \neq N - 1$, then at least one of the two numbers given by

$$\gcd(N, a^{r/2} \pm 1) \tag{9.15}$$

is a non-trivial factor of N. If r does not have the required properties then one can simply pick another value of a and try again. In fact, it turns out that the process works for the majority of possible values of a, and so Shor's algorithm is very likely to work after a few tries.

9.5 Grover's algorithm

The algorithms described above are all very similar: they all begin by evaluating some function over a superposition of all possible inputs, and then use some clever transformation of the output values to extract some desired result which summarizes all the values. Formally speaking these algorithms tackle the *Abelian hidden subgroup* family of problems, based on identifying some hidden structure in the pattern of function values. By contrast, Grover's algorithm seeks to identify *particular inputs* with interesting output values.

In its simplest form, described briefly in the previous chapter, Grover's algorithm seeks to identify a unique satisfying input value, for which $f(x) = 1$, from amongst $N = 2^n$ possible inputs, the rest of which have $f(x) = 0$. The case $n = 2$, where we are seeking for one satisfying input among four possibilities, is particularly simple, and can be understood using brute-force methods, but to understand larger cases it is better to use a more sophisticated treatment.

As usual, Grover's algorithm is assumed to have access to an oracle implementation of f which performs

$$|x\rangle|y\rangle \xrightarrow{U_f} |x\rangle|y \oplus f(x)\rangle, \tag{9.16}$$

where $|y\rangle$ is a single ancilla qubit. Using the usual phase kickback trick this can in effect be replaced by a purely quantum oracle, which performs the transformation

$$V|x\rangle = (-1)^{f(x)}|x\rangle, \tag{9.17}$$

simply returning $|x\rangle$ unchanged unless it is the unique satisfying value $|a\rangle$, in which case it is converted to $-|a\rangle$. We can describe V in another way by thinking about its action on a general superposition state $|\psi\rangle$; this can be expanded in the computational basis as

$$|\psi\rangle = \mathbb{1}|\psi\rangle = \sum_x |x\rangle\langle x|\psi\rangle \tag{9.18}$$

and the effect of V is simply to negate the component of $|\psi\rangle$ along $|a\rangle$. Thus we can write

$$V_a|\psi\rangle = |\psi\rangle - 2|a\rangle\langle a|\psi\rangle \tag{9.19}$$

or

$$V_a = \mathbb{1} - 2|a\rangle\langle a|. \tag{9.20}$$

Grover's quantum search begins by applying the quantum oracle V_a to a uniform superposition of states

$$|\phi\rangle = H^{\otimes n}|0\dots00\rangle, \tag{9.21}$$

which for the case $n = 2$ is accomplished by the first three columns of gates in the network depicted in equation (8.19). The next three columns are intended to implement the operation

$$W = 2|\phi\rangle\langle\phi| - \mathbb{1} = H^{\otimes n}(-V_0)H^{\otimes n}. \tag{9.22}$$

Note that as drawn, the network actually implements $-W$ rather than W, but as these differ only by a global phase this distinction is unimportant. The actions of these operators can be summarized as

$$W|\phi\rangle = |\phi\rangle \qquad W|a\rangle = \frac{2}{\sqrt{N}}|\phi\rangle - |a\rangle \tag{9.23}$$

and

$$V|\phi\rangle = |\phi\rangle - \frac{2}{\sqrt{N}}|a\rangle \qquad V|a\rangle = -|a\rangle, \tag{9.24}$$

where we have used the fact that $\langle a|\phi\rangle = 1/\sqrt{N}$, as $|\phi\rangle$ is a uniform superposition of all possible states.

For the case $N = 4$ the analysis is simple. Starting from the state $|\phi\rangle$ the function evaluation operator produces

$$V|\phi\rangle = |\phi\rangle - |a\rangle, \tag{9.25}$$

in which the amplitude of the desired state has been labeled with a minus sign (recall that $|\phi\rangle$ includes a component $\frac{1}{2}|a\rangle$, which is converted to $-\frac{1}{2}|a\rangle$ by the operator V). Next the amplitude amplification operator W acts to give

$$W(|\phi\rangle - |a\rangle) = W|\phi\rangle - W|a\rangle = |\phi\rangle - (|\phi\rangle - |a\rangle) = |a\rangle, \tag{9.26}$$

and so the final state of the computer is the desired satisfying state. For larger values of N, however, this process moves the state closer to $|a\rangle$, but does not take it all the way there. Instead, it is necessary to repeatedly apply V and W alternately to the state, moving it further toward $|a\rangle$ at each stage. To see how this works we need a simple way of evaluating $(WV)^r$, which describes the effect of this repeated evolution.

The key feature of equations (9.23) and (9.24) is that V and W cause the two vectors $|a\rangle$ and $|\phi\rangle$ to evolve within a subspace defined by these two vectors. These vectors, however, do not form a good basis for a subspace as they are not orthogonal. This can be resolved by replacing $|\phi\rangle$ with

$$|\phi_0\rangle = \frac{\sqrt{N}|\phi\rangle - |a\rangle}{\sqrt{N-1}}, \tag{9.27}$$

which can be combined with $|a\rangle$ to form an orthonormal basis. Note that $|\phi_0\rangle$ is a uniform superposition of all the inputs which do not satisfy f, and we can write

$$|\phi\rangle = \frac{\sqrt{N-1}|\phi_0\rangle + |a\rangle}{\sqrt{N}}. \tag{9.28}$$

Using the $\{|\phi_0\rangle, |a\rangle\}$ basis the two operators now take the simple forms

$$V = \begin{pmatrix} 1 & 0 \\ 0 & -1 \end{pmatrix} \qquad W = \begin{pmatrix} \frac{N-2}{N} & \frac{2\sqrt{N-1}}{N} \\ \frac{2\sqrt{N-1}}{N} & -\frac{N-2}{N} \end{pmatrix} \tag{9.29}$$

and the combined effect of the operators is

$$WV = \begin{pmatrix} \frac{N-2}{N} & -\frac{2\sqrt{N-1}}{N} \\ \frac{2\sqrt{N-1}}{N} & \frac{N-2}{N} \end{pmatrix} = \begin{pmatrix} \cos\theta & -\sin\theta \\ \sin\theta & \cos\theta \end{pmatrix} = \exp(-\mathrm{i}2\theta\,\sigma_y/2), \tag{9.30}$$

which has the form of a rotation of the Bloch sphere in this subspace, with

$$\theta = \arctan\left(\frac{2\sqrt{N-1}}{N-2}\right) \approx \frac{2}{\sqrt{N}} \tag{9.31}$$

where the approximation applies in the limit that N is very large.

The behavior of Grover's algorithm in the case of large N is now clear. The first set of Hadamard gates produces the state $|\phi\rangle$, which for large N is almost identical to $|\phi_0\rangle$. The effect of applying the operator WV is to rotate this state by an angle 2θ toward the desired state $|a\rangle$, and if the operator is applied repeatedly the state will initially rotate further toward it. After r repetitions of V and W the total rotation angle will be $2r\theta$, and when this angle is close to π then the state will be essentially equal to the desired satisfying state $|a\rangle$. Solving for $2r\theta = \pi$ gives the required number of repetitions as

$$r \approx \frac{\pi}{4}\sqrt{N}, \tag{9.32}$$

showing the expected reduction in function evaluations from around $N/2$ to about $\sqrt{N}$. Equivalently, since $N = 2^n$, where n is the number of input bits, the number of evaluations is reduced from 2^{n-1} to $2^{n/2}$.

If N is not very large then this approximate calculation must be corrected in two ways. Firstly the exact expression for θ in equation (9.31) must be used, and secondly the correct initial state must be used. The initial state $|\phi\rangle$ is not quite $|\phi_0\rangle$, but is in fact slightly rotated toward $|a\rangle$ by some small angle θ_0, and so the total rotation angle required is slightly less than π. The exact solution is

$$r = \frac{\pi - 2\arctan[1/\sqrt{N-1}]}{2\arctan[2\sqrt{N-1}/(N-2)]}, \tag{9.33}$$

which reduces to equation (9.32) in the limit of large N; as expected, this gives the exact result $r = 1$ in the simple case $N = 4$.

In general, equation (9.33) does not have integer solutions for r, and so there is no number of repetitions which will exactly produce the desired final state $|a\rangle$. However, if the number of repetitions is close to r then the final state will be close to $|a\rangle$, and measuring it will give the result $|a\rangle$ with high probability. The result obtained can easily be checked by evaluating $f(x)$, and if x is not in fact a satisfying input then the algorithm is simply repeated; it is very unlikely that the algorithm will give an erroneous result a second time. The naive approach of choosing the nearest integer to equation (9.32) for the number of repetitions gives an algorithm that works with more than 95% reliability if n is at least four, and with more than 99% reliability if n is at least nine. If the number of repetitions is further increased beyond this ideal value then the probability of finding the satisfying input decreases, as the state is now driven away from $|a\rangle$, back toward $|\phi_0\rangle$.

9.6 Generalizing Grover's algorithm

This approach also reveals two simple generalizations of Grover's algorithm, which apply in the case where the function has more than one satisfying input. The analysis is remarkably similar to that used in the previous section, except that the basis vectors used to define the subspace are now $|\phi_1\rangle$, a uniform superposition of the k satisfying inputs, and $|\phi_0\rangle$, a uniform superposition of the $N - k$ remaining inputs. The overall result is also very similar, and in the limit that N is very large in comparison with k the final state is very close to $|\phi_1\rangle$ after

$$r_k \approx \frac{\pi}{4}\sqrt{N/k} \tag{9.34}$$

repetitions. Measuring the final state will then return *one* of the satisfying inputs, chosen at random.

Example 9.4 The simplest case to consider is when $k = N/4$, as in this case the situation is equivalent to the case of a single satisfying input from among four possibilities. In this case it is relatively simple to work through the algorithm by hand, starting from the state

$$|\phi\rangle = \frac{\sqrt{3}|\phi_0\rangle + |\phi_1\rangle}{2} \tag{9.35}$$

and applying first $V = \mathbb{1} - 2|\phi_1\rangle\langle\phi_1|$ and then $W = 2|\phi\rangle\langle\phi| - \mathbb{1}$ to give the final state $|\phi_1\rangle$.

If the value of k is known then this generalization of Grover's algorithm is, in effect, identical to the standard version, but a problem arises if k is unknown, as in this case it is impossible to choose an optimal number of repetitions without first having a reasonable estimate of k. This can be achieved by using a variant of Grover's algorithm called *approximate quantum counting*.

Quantum counting relies on the fact that the repeated operation WV is a rotation of the form $(2\theta)_y$ in the subspace defined by $|\phi_0\rangle$ and $|\phi_1\rangle$, and there is a simple algorithm for estimating the angle of a rotation operator. Consider the effect of the network

$$\tag{9.36}$$

where the central gate corresponds to applying a controlled-$(2\theta)_y$ rotation r times. This gate is applied to the state $|+\rangle|0\rangle$ to give

$$\begin{pmatrix} 1 & 0 & 0 & 0 \\ 0 & 1 & 0 & 0 \\ 0 & 0 & \cos r\theta & -\sin r\theta \\ 0 & 0 & \sin r\theta & \cos r\theta \end{pmatrix} \begin{pmatrix} \frac{1}{\sqrt{2}} \\ 0 \\ \frac{1}{\sqrt{2}} \\ 0 \end{pmatrix} = \frac{1}{\sqrt{2}} \begin{pmatrix} 1 \\ 0 \\ \cos r\theta \\ \sin r\theta \end{pmatrix}. \tag{9.37}$$

Applying the final Hadamard gate to the first (control) qubit converts this to

$$\frac{1}{2} \begin{pmatrix} 1 + \cos r\theta \\ \sin r\theta \\ 1 - \cos r\theta \\ -\sin r\theta \end{pmatrix}, \tag{9.38}$$

in which the state of the control qubit now depends on $r\theta$. This is clearly revealed by measuring the first qubit in the computational basis, which can give either $|0\rangle$ with probability

$$P(0) = \frac{(1 + \cos r\theta)^2 + (\sin r\theta)^2}{4} = \cos^2(r\theta/2) \tag{9.39}$$

or $|1\rangle$ with probability $P(1) = \sin^2(r\theta/2)$. Although the result of any one measurement is random, repeating the network several times allows $r\theta$ to be estimated, and varying r allows θ to be estimated. This network can be generalized in several ways.

Example 9.5 Repeating this calculation with the second qubit in the general state $\alpha|0\rangle + \beta|1\rangle$ shows that the network gives the same result whatever the initial state of this qubit.

Since the Grover operator WV corresponds to a rotation in the two-dimensional subspace defined by $|\phi_0\rangle$ and $|\phi_1\rangle$ with a rotation angle

$$\theta(k, N) \approx \frac{2}{\sqrt{N/k}}, \tag{9.40}$$

a network of this kind can be used to estimate k. Note, however, that this requires the ability to implement the n-qubit Grover network conditionally, controlled by an additional qubit in state $|+\rangle$. There are several sophisticated variations of this approach, including Grover's *fixed-point quantum search*, which converges on the desired state for *any* value of k, but these algorithms are beyond the scope of this text.

9.7 Quantum simulation

In addition to the period-finding algorithms (such as Deutsch and Shor) and the search algorithms (such as Grover), there is a third significant group of quantum algorithms based on *quantum simulation*. The basic idea is to use a quantum computer to simulate the behavior of another quantum system where it is easy to write down the Hamiltonian, but hard to calculate the consequences. For example, it might be possible to investigate various models of superconductivity by determining whether these models correctly predict a range of physical phenomena. This may well turn out to be the most important application of quantum computers in real life.

The core idea of quantum simulations is fairly straightforward. Any physical system of interest will have a Hamiltonian, and the evolution of the system under this Hamiltonian for some fixed time t will be a unitary propagator. Since quantum logic gates are universal, any desired evolution can be implemented with some suitable network of gates; in the same way the initial state of the system can be encoded as some state of a set of qubits, and the final state of the system can then be calculated by evolving the initial state under the quantum network. The simplest approach is just to calculate the evolution of the system directly, but using techniques similar to those behind the quantum counting algorithm it is possible to estimate more complex properties, such as eigenvalues of the underlying Hamiltonian.

This process will only have an advantage over conventional computation if it is possible to design the evolution network without explicitly evaluating the propagator for the quantum system being simulated. In effect this means that it is necessary to encode not the propagator, but rather the Hamiltonian of interest. The basic route for doing this relies on a set of remarkable formulae for approximating the evolution of a system under the sum of two non-commuting Hamiltonians.

Suppose the Hamiltonian of the system being simulated can be written in the simple form $\mathcal{H} = \mathcal{A} + \mathcal{B}$, where the two sub-Hamiltonians can be implemented directly. This does not solve the problem, as the evolution under the sum cannot be directly evaluated by combining evolutions under the two parts: that is

$$e^{-i(\mathcal{A}+\mathcal{B})t/\hbar} \neq e^{-i\mathcal{A}t/\hbar}e^{-i\mathcal{B}t/\hbar} \tag{9.41}$$

unless $\mathcal{A}$ and $\mathcal{B}$ commute, so that $[\mathcal{A}, \mathcal{B}] = 0$. However, the related equation

$$e^{-i(\mathcal{A}+\mathcal{B})t/\hbar} = \lim_{n\to\infty} \left(e^{-i\mathcal{A}t/n\hbar}e^{-i\mathcal{B}t/n\hbar}\right)^n \tag{9.42}$$

holds even if $[\mathcal{A}, \mathcal{B}] \neq 0$. In essence this equation reflects the fact that for sufficiently large n, the evolution time t/n is so short that *all* the operators on the right-hand side are very near to the identity operator, and all such operators very nearly commute.

In principle this approach is sufficient to solve the problem, but in practice it may be necessary to use very large values of n to get reasonably stable behavior. However, methods exist by which more rapid convergence can be achieved. In particular, it can be shown that

$$e^{-i(\mathcal{A}+\mathcal{B})\delta t/\hbar} = e^{-i\mathcal{A}\delta t/\hbar}e^{-i\mathcal{B}\delta t/\hbar} + O(\delta t^2) \tag{9.43}$$

while

$$e^{-i(\mathcal{A}+\mathcal{B})\delta t/\hbar} = e^{-i\mathcal{A}\delta t/2\hbar}e^{-i\mathcal{B}\delta t/\hbar}e^{-i\mathcal{A}\delta t/2\hbar} + O(\delta t^3) \tag{9.44}$$

gives more rapid convergence. Explicit networks have been developed for simulating a variety of important physical systems, but the details of these are beyond the scope of this text.

9.8 Experimental implementations

In the next two chapters we will consider two approaches for implementing quantum computing, firstly based on energy levels in atoms or ions, and secondly based on nuclear spins. Methods for implementing single-qubit gates in these systems were discussed in Part I, and we now consider two-qubit gates, as well as taking a harder look at experimental practicalities. As usual we will structure the discussion around the first five DiVincenzo criteria, except that the question of scalability (that is, whether the technique can be scaled up to produce a large-scale general-purpose computer) will be left until later.

Further reading

These more advanced quantum algorithms are also covered in most standard texts, but Mermin (2007) provides a particularly good introduction. The Hadamard transform is rarely discussed in physics texts outside the context of quantum information, but is discussed in books on signal processing such as Beauchamp (1987). The original description of Shor's quantum factoring algorithm is hard to locate; a good review was subsequently published (Shor, 1999), but this remains a difficult algorithm to understand at an elementary level.

A more thorough discussion of public key algorithms which may not be vulnerable to quantum computers can be found in Bernstein *et al.* (2010). A brief discussion of quantum simulation, and in particular of the Trotter formulae described above, can be found in Nielsen and Chuang (2000).

Exercises

9.1   Consider a Deutsch–Jozsa problem with $n = 2$: how many possible functions are there, and how many are constant and how many are balanced? Assuming that an unknown function is known to be either constant or balanced with 50% probability, calculate the minimum, maximum, and average number of queries required to determine which sort of function it is on a classical computer. What about a quantum computer?

9.2   Consider Grover's algorithm in the case $N = 16$ and $k = 1$. Calculate the exact value of the rotation angle θ and the initial angle θ_0, and hence find the probability of ending up in the desired satisfying state for values of r between 0 and 3 inclusive.

9.3   Compare the results of the previous calculation with those obtained from the approximate solutions at large N, namely $\theta = 2/\sqrt{N}$ and $\theta_0 = 0$. Why does equation (9.32) work well even at relatively small values of N?

9.4   Prove the result concerning the quantum counting algorithm asserted in Example 9.5. Use this result to show that this network would work even if the second qubit starts in the maximally mixed state.

9.5   Show that equations (9.43) and (9.44) are correct for the case $\mathcal{A} = \hbar \sigma_z$ and $\mathcal{B} = \hbar \sigma_x$.

10 Trapped atoms and ions

As discussed in Part I of this book, a qubit can in principle be encoded using two energy levels in an atom or ion. These levels are frequently referred to as $|g\rangle$ and $|e\rangle$, suggesting the ground state and some excited state, but the actual choice of levels is made so as to optimize the behavior of the system, and it is common to use two hyperfine sublevels of the ground state. As usual, we will simply call the levels $|0\rangle$ and $|1\rangle$. A quantum computer must, of course, have more than one qubit, and this is achieved by using more than one atom or ion, with one qubit encoded on each physical object. In order to make this approach practical, however, it is essential to trap the atoms or ions so that they can be held in a well-controlled environment where they can easily be manipulated. This can be achieved using electric and magnetic fields.

The use of trapped ions was one of the first proposals for building quantum computers, and is still one of the best developed. Proposals involving trapped atoms are slightly more recent, and have both substantial advantages and disadvantages in comparison with trapped ions. These differences can ultimately be traced back to the fact that ions interact strongly with their environment and have long-range interactions with one another, while atoms interact more weakly and over shorter ranges. Comparing and contrasting the two approaches provides a useful general introduction to the problems underlying many other proposals.

10.1 Ion traps

It is relatively easy to trap an ion as it will interact strongly with an electric field through the Coulomb interaction. Here we will assume that the ion is positively charged; negatively charged ions can of course be trapped in a very similar way. An obvious first idea about how to build a trap is simply to surround the ion with positively charged electrodes, each of which will repel it, so that it remains in the center of the system. A little thought, however, reveals that this process cannot be made to work: a trapped ion must sit at a minimum of the electrostatic potential produced by the electrodes, and the existence of such a minimum in free space would violate Gauss's law.

For example, consider an arrangement of six positive charges, placed at equal distances on the $\pm x$, $\pm y$ and $\pm z$ axes from the ion, forming an octahedron. For motion *along* the axes the potential increases as the ion moves from the origin, and so it seems that this potential will confine the ion. For motion *between* the axes, however, the potential decreases, and so the ion will be expelled in one of these directions. (There is a metastable equilibrium point at the very center of the trap, but if the ion wanders from the exact center it will eventually be expelled from the trap.) Another possibility to consider is surrounding the ion with a uniform sphere of positive charge. Naively one might guess that the ion would be repelled by the charges toward the center of the sphere, but elementary electrostatics shows that the electric potential inside a uniformly charged sphere is in fact completely uniform, and thus the ion will feel no force at all.

It therefore appears that ion trapping is impossible! Fortunately this is not the case, and one answer is to replace the *static* arrangement of charges with a time-varying electric field. This can be done in such a way that the time-averaged potential does indeed have a minimum at the center of the trap. A simple analogy can be made with the situation of a ball sitting on a saddle-shaped surface. Clearly the ball is trapped in one direction but repelled in the other, and left alone the ball will simply fall off the saddle. If, however, the saddle is made to rotate at an appropriate angular velocity then the rising edge of the saddle will "catch up" with the ball as it begins to fall, so that the ball remains trapped. While the *instantaneous* electrostatic potential has repulsive directions, the time-averaged potential can confine the ion. As before, ions at the exact center are now stably trapped, but ions a small distance from the center will no longer be expelled from the trap but will travel around the center, effectively surfing the rotating potential. (This circular *micromotion* is potentially annoying, but can largely be suppressed by forcing the ions toward the center of the trap.)

The mathematics of the situation is similar to that found in the linear Paul trap, a common design used for ion trap quantum computing. This uses a rotating electrical potential, obtained by applying oscillating voltages to quadrupolar electrodes, to produce a confining potential which is strong in two directions (x and y) and weak in the third direction (z). If two or more ions are placed in the trap they will repel one another through their mutual Coulomb interactions, and the final result will be a linear string of ions arranged along the z axis. The spacing between the ions can be controlled by varying the strength of the trap along this axis, and is typically around 10 μm. An alternative solution, which is adopted in the Penning trap, combines electrostatic and magnetic fields, but we do not consider this further.

So far we have assumed that the only effect of the trapping potential is simply to keep the ion (or ions) in one place, but the real situation is much more complex than this. As the ion is confined, its motional states become quantized. This can usually be approximated by a harmonic oscillator potential, and so the energy levels of the trapped ion are replaced by ladders of energy levels. This point is considered in more detail in Section 10.3.

10.2 Atom traps and optical lattices

Atoms, unlike ions, are uncharged, and so cannot be trapped with electric fields: instead, they are trapped using light. The most obvious approach is based on the *scattering force* which occurs when atoms absorb photons coming from one direction and re-emit them at random, resulting in a net change of momentum. When applied to atoms in an atomic beam, which are initially traveling in similar directions at similar speeds, this approach can be used to bring atoms almost to a halt. A more sophisticated approach, known as *optical molasses* uses six beams, one directed along each axis and all tuned slightly below a transition. The Doppler effect means that atoms will preferentially absorb photons from the beam toward which they are traveling (these photons are blue-shifted toward resonance), and so atoms are effectively prevented from moving rapidly in any direction. This leads to a reduction in the atomic velocities, which is usually described as *cooling* the atoms, but while low temperatures can be reached this approach is limited by the *Doppler cooling limit*, which arises in essence from the fact that there is a minimum momentum transfer corresponding to the absorption of a single photon. Fortunately this limit can be surpassed by more complex sub-Doppler cooling techniques, and further extended with evaporative cooling, ultimately leading to an ultracold *Bose–Einstein condensation*, or BEC. An even more powerful trap, known as a *magneto optical trap* or MOT, can be obtained by combining magnetic fields with circularly polarized laser beams. This allows large numbers of cold atoms to be stored for later use.

A more subtle form of optical trapping is based on the *dipole force*. The basic mechanism can easily be understood by considering the forces on a prism which refracts a light beam. As the light beam is bent, its momentum is changed, and so there must be a corresponding change in the momentum of the prism. Thus the prism feels a force, pushing it in the opposite direction to the light beam. This force will depend on both the angle of the incoming light (as well as its intensity of course) and on the refractive index of the prism. A similar situation occurs with an atom in a light field, except (of course) that an atom is not shaped like a prism. A better model is to treat the atom as a spherical lens, either focusing or defocusing the light. If the light is of uniform intensity then there is no overall force on the sphere, but if the intensity varies across the sphere then it will feel a force that depends on the gradient of the intensity. Depending on the relative refractive index of the sphere, η_s, and the medium, η_m, the sphere will be pushed toward the region of highest light intensity (if $\eta_s > \eta_m$) or the region of lowest light intensity (if $\eta_s < \eta_m$). As the refractive index of a material changes substantially close to an absorption line, the properties of the force will depend on the relative frequency of the light and that of the relevant transition. This is the basis of *optical tweezers*, which have found extensive application in biophysics and nanotechnology.

This description is not really applicable to atoms, as refractive index is a property of bulk materials, but a proper mathematical treatment leads to very similar results. The electric field of light can induce an oscillating dipole in an atom, which then interacts with the light field (hence the name of dipole force). The magnitude of this force is greatest when the frequency of the light is close to resonance, and its direction depends on whether the light is tuned above or below the transition. In a spatially varying light field, atoms will seek either the regions of highest intensity or those of lowest intensity, depending on the frequency of the light, allowing traps to be constructed. This idea underlies the use of *optical lattices* to manipulate very large numbers of atoms in equally large numbers of traps. A standing wave laser field is created, which gives a light field whose intensity varies periodically with a period equal to half the wavelength of the light. It turns out that the dipole force is particularly effective in this arrangement, allowing atoms to be trapped in each well. With

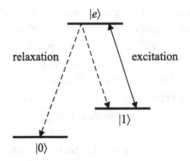

Optical pumping. The excited state $|e\rangle$ relaxes down to the two states $|0\rangle$ and $|1\rangle$ at random, but as the undesired state $|1\rangle$ is continuously pumped up to $|e\rangle$ while the desired state $|0\rangle$ is left alone, eventually all the population will be transferred to $|0\rangle$.

a little more effort a two (or three)-dimensional standing wave can be created, giving a two (or three)-dimensional array of these microtraps, sometimes described as an "egg box for atoms."

The traps in these optical lattices are not very deep, and so can only trap cold atoms. This is achieved by loading them with atoms from a MOT or from a previously prepared BEC. (The advantage of using a BEC is that the atoms are so cold that they end up in the lowest vibrational band of the lattice, which arises due to quantum tunneling between different traps.) These atoms are initially delocalized across the entire lattice, but on raising the light intensity, and thus making the microtraps deeper, the atoms become localized on individual lattice sites, so that exactly one atom ends up in each trap. The resulting array of single atoms is known as a Mott-insulator state, and their internal states can then be used as qubits.

10.3 Initialization

Having trapped the atoms or ions it is necessary to ensure that their qubits are all in some well-defined initial state, usually $|0\rangle$, before they can be used for a computation. If the two basis states were indeed a ground and excited state, then this could be achieved by direct cooling, but in practice more subtle mechanisms are required, and this is achieved through optical pumping. The basic idea is very simple: a laser is used to excite atoms which are in any state other than the desired initial state into some high-energy state $|e\rangle$, and combined with random relaxation processes this leads to preferential population of the desired state, as depicted in Figure 10.1.

The excitation process can be thought of as a series of Rabi π pulses, which coherently transfer all the population from the low-energy state to the excited state, separated by relaxation periods. These pulses would also transfer population back down, but due to rapid relaxation of the excited state there is no population to transfer down. Alternatively, the excitation can be achieved with low-power continuous irradiation, in which case the

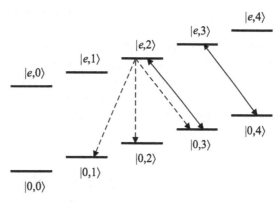

Fig. 10.2   Sideband cooling by optical pumping. A single laser can excite all the $\Delta n = -1$ transitions within the vibrational ladder of the $|0, n\rangle \leftrightarrow |e, n'\rangle$ transition, with subsequent relaxation distributing the population among the lower levels. As there are no $\Delta n = -1$ transitions from the lowest $|0, 0\rangle$ level, all the population is eventually pumped to this state.

population transfer and relaxation occur in parallel, with both processes being described by rate equations.

With trapped ions it is important to note that in addition to cooling the qubit itself, it is also important to cool the vibrational modes of the ions within the trap potential. We can label the state of an ion by giving both the qubit state and the vibrational state; for example, the state $|0, n\rangle$ indicates a qubit in state $|0\rangle$ in an atom in vibrational state n. As vibrational excitations are frequently used to implement quantum logic gates it is necessary to begin with every atom in its vibrational ground state, and so it is useful to be able to prepare the state $|0, 0\rangle$. This can also be achieved by optical pumping, a technique known as sideband cooling depicted in Figure 10.2. Sideband cooling cannot be used with trapped atoms, as the momentum kick of a single emitted photon is sufficient to destroy the whole array; however, such cooling is unnecessary as the process begins with an ultracold BEC.

10.4 Decoherence

Decoherence processes occur in any physical system, and are the great enemy of quantum information processing as they cause coherent superpositions to decay into classical mixtures. In essence, all decoherence can be traced back to uncontrolled interactions with the environment. A detailed treatment is well beyond the scope of this text, but it is immediately obvious that long-range Coulomb forces may give rise to serious problems in trapped ions, and ion traps are particularly vulnerable to fluctuations in the potentials of the confining electrodes and to effects of charges trapped on nearby insulators.

Tackling decoherence is also the fundamental reason for using two hyperfine levels within the ground state, rather than a ground state and an excited state. The ultimate limit to decoherence is provided by the spontaneous emission lifetime of a transition, and this decreases very rapidly for the high-energy transitions to excited states. The detailed choice of two individual hyperfine levels in trapped ions usually depends on the dominant decoherence mechanisms; for example, it may be advantageous to pick two levels which are only weakly affected by fluctuating magnetic fields. With trapped atoms the choice of levels is partly determined by the details of the transition used to implement the trapping potential, but where it is possible picking levels which are unaffected by stray fields, or which are both affected in the same way, remains a sensible approach.

10.5 Universal logic

Clearly it is essential to be able to perform universal quantum logic if interesting computations are to be achieved. As mentioned previously, it can be shown that the combination of the controlled-NOT gate and a small set of single-qubit gates is universal for quantum information processing, meaning that any desired operation can be built from a network of these gates; in particular, three-qubit gates (such as the Toffoli gate) are not required. This is a key result in experimental quantum information processing as directly implementing a three-qubit gate would require a physical interaction involving three particles: fortunately, we only require interactions involving one or two atoms or ions.

Single-qubit gates are (in principle) simple for trapped ions, as these can be achieved using Raman transitions induced by shining lasers on the ion of interest. By controlling the power, duration, and phase of the laser pulse a wide range of different single-qubit rotations can be directly achieved, and any remaining single-qubit gates required can be constructed out of networks of these basic gates. Of course, single-qubit gates require that only one qubit experiences the rotation, but this is relatively simple as the spacing between ions is large compared to the wavelength of the laser light used, and so in principle it is not too difficult to focus the lasers down onto single ions.

The situation with atoms trapped in optical lattices is more problematic. There is no problem in using Raman transitions to induce rotations, but selectively exciting a single qubit initially appears very difficult. The separation between individual atoms is usually only half a wavelength of the light used to set up the lattice, making it extremely difficult to focus on a single atom. Until recently this apparent inability to implement single-qubit gates selectively was a major problem with trapped atoms. It is, however, easy to carry out simultaneous identical single-qubit gates on very large numbers of trapped atoms, and this intrinsic massive parallelism is a topic of considerable interest in optical lattice research.

The most obvious approach to implementing selective single-qubit gates with trapped atoms is to increase the separation between atoms by only sparsely filling the lattice of traps. With small one-dimensional systems random sparse filling can create small regions with suitable patterns of occupation, but this is not realistic in larger systems. A more systematic approach is to use an electron beam (which can be very tightly focused) to depopulate selected sites in the optical lattice, leaving atoms in only a small number of desired locations. However, neither of these approaches is entirely satisfactory, and it would be preferable to find some way of directly addressing individual atoms on adjacent lattice sites. (Resolving individual atoms for selective readout is considered separately below.)

This has now been achieved by using a tightly focused laser beam not to induce transitions, but rather to shift the energy levels within atoms by the AC Stark shift. If a differential shift can be implemented for the levels $|0\rangle$ and $|1\rangle$, then the resonance frequency of the transition will depend on the intensity of the light inducing the shift; this will be higher at the very center of the laser spot than at its edges, and so the transition frequency of the central atom will be significantly different from all its neighbors. Direct (rather than Raman) transitions can then be selectively implemented using a microwave source which is resonant with the central atom frequency, but not with any of the surrounding transitions. As the selectivity is achieved using different resonance frequencies for different atoms there is no need to focus this microwave beam, which can uniformly illuminate the whole lattice.

Next it is necessary to implement a two-qubit gate, such as the controlled-NOT gate. It is important to note that the controlled-NOT gate is not the only universal two-qubit gate, and in fact any non-trivial two-qubit gate, and thus almost any physical interaction between the atoms or ions carrying the qubit, can be used as the basis of universal logic. (To make this statement useful it is necessary to give a better definition of *trivial*, or equivalently *non-trivial*. One simple approach is to note that any gate which can convert a product state into an entangled state is non-trivial.) The natural two-qubit gate is different for atoms and ions, reflecting the different physical interactions used.

10.6   Two-qubit gates with ions

For trapped ions it is simple to explain how a gate can be built in principle, although the details of actually doing it in practice are quite complex. The basic interaction used is the Coulomb repulsion between ions, which links the motional degrees of freedom of the ions into common vibrational modes. To put it crudely, if one ion in a trap is waggled the others will certainly notice, and will attempt to follow its motion. This common vibrational mode acts as a "data bus," carrying quantum information between different ions and so between different qubits.

The basic idea is to use selective Raman transitions which excite motions in the first ion if and only if it is in the (qubit) excited state, in effect transferring the information stored in the superposition from the qubit state into the vibrational state of the ion. This can be achieved by applying a Rabi π pulse focused on ion 1 and tuned to the transition $|1, 0\rangle \leftrightarrow |0, 1\rangle$, as shown in Figure 10.3. Note that for ions in the state $|0, 0\rangle$ there is no resonant transition which can be excited; there are, of course, available transitions for ions in higher vibrational levels, but as the ions have initially been cooled to their vibrational ground state there are no ions in these levels. The result of this process is to transfer the quantum information from the internal state of the first ion to its vibrational state.

Since the motion of all the ions is coupled, this information is now available at each and every other ion in the form of its own vibrational state. Hence if we apply an operation to ion 2 conditional on its vibrational state then we have applied this same operation conditional

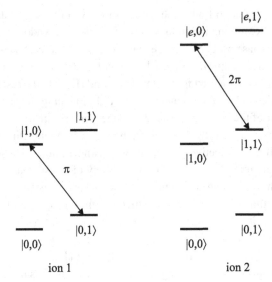

Fig. 10.3 Transitions used to implement a two-qubit logic gate with trapped ions; as usual the notation $|q, n\rangle$ indicates a qubit in state $|q\rangle$ in an atom in vibrational level n. For details see the main text.

on the original internal state of ion 1. The operation chosen is a Rabi 2π pulse tuned to the transition $|1, 1\rangle \leftrightarrow |e, 0\rangle$ involving some excited state $|e\rangle$. This 2π rotation might seem to have no effect, but in fact the quantum state picks up a phase of -1, reflecting the spinor behavior of qubits described in Section 1.3. Finally a second Rabi π pulse applied to the first ion transfers its quantum information back to its internal states, and restores the whole system to the vibrational ground state. The overall process is easily seen to be a controlled-Z gate between the two ions, which is easily converted to a controlled-NOT gate by the usual network.

10.7 Two-qubit gates with atoms

This approach cannot be used with trapped atoms, as they do not have the long-range Coulomb interactions required. It is possible to use dipole–dipole interactions between magnetic dipoles or induced electric dipoles, but here we concentrate on a more common approach, based on the contact or collision interaction.

Although atoms do not normally interact strongly at long distances, they repel each other very strongly at short distances, so that it is not possible to squeeze two atoms into the space normally occupied by one. The microtraps in optical lattices are so small that this effect can be quite significant, and so the energy of an atom in a trap will depend on whether or not there is another atom in the same trap (Figure 10.4). In general, the act of bringing two atoms closer together will raise their energy, and therefore (by the time-dependent Schrödinger equation) cause them to pick up an additional phase shift, beyond that which they would

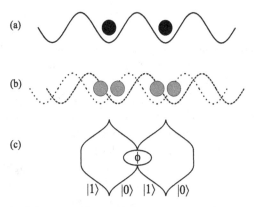

Fig. 10.4 (a) Atoms sit at the bottom of wells in an optical lattice. (b) If the lattice is made from two separate lattices produced with circularly polarized light then the two lattices can be moved apart. Which lattice the atom feels depends on its internal state, and thus each atom's wavepacket is split into two spatially separated components. (c) Atoms in neighboring wells can be brought into contact, raising their energies and so applying a phase shift to the system; this will only occur if the two atoms have appropriate internal states.

have acquired if left alone. Note that to a good approximation only the *energies* of the atoms are changed, and not their wavefunctions, as expected from first-order perturbation theory.

Detailed calculations of the strength of the contact interaction are quite complex, but the key results are fairly simple. As the interatomic potential is usually short range only head-on collisions matter, and so the collision can be described using s-wave scattering, in which the relative orbital angular momentum of the colliding particles is assumed to be zero. In this case the interaction turns out to be describable by a single parameter, the scattering length a. The energy shift arising from the collisional interaction between two identical atoms of mass M is then approximately given by

$$E_a = \frac{4\pi\hbar^2 a}{M} \int |\psi|^4 \, d\mathbf{r}, \qquad (10.1)$$

where ψ is the vibrational wavefunction of the atom in the trap. The fourth-power dependence on ψ corresponds to a quadratic dependence on the probability of the atom being found at some point in the well, and so the integral gives the probability of two atoms being found at the same point in space.[1] Clearly this depends strongly on the shape of the trapping potential. Note that although the scattering length has the dimensions of length it does not simply correspond to the atomic size: it can vary greatly with fine details of the situation, can be controlled by applying external fields, and can even be negative.

If two atoms are made to collide for a time τ, the system picks up an additional phase shift $e^{-i\phi}$, with $\phi = E_a \tau / \hbar$. To implement a controlled logic gate with this interaction it is necessary to make the additional phase shift depend on the qubit state of the two atoms involved. This can be achieved by making the trapping potential depend on the qubit

---

[1] This calculation assumes that the two atoms have been brought to the same location in space, so that their individual wavefunctions are identical and fully overlapping. A slightly more complex expression is necessary in more general cases.

state, so that atoms in state $|0\rangle$ will feel one potential, while atoms in state $|1\rangle$ will feel a different potential. If these potentials can be controlled independently, then atoms can be moved in different ways depending on their qubit state. Since this whole process is quantum mechanical, an atom whose qubit is in a coherent superposition of both $|0\rangle$ and $|1\rangle$ will move in both directions at once!

This effect can be achieved when the qubit corresponds to the $m_S = \pm\frac{1}{2}$ spin states of the $S_{1/2}$ ground state of an atom such as Rb. (In reality it is necessary to include the effects of nuclear spin and select two appropriate hyperfine states, but the essentials of the situation remain the same.) In this case the laser frequency can be chosen such that the $\pm\frac{1}{2}$ spin states interact with the $\sigma^{\mp}$ components of the standing light wave, and these can be controlled by varying the phase of the light. In particular, it is possible to arrange that the σ^- traps, and thus the $m_S = +\frac{1}{2}$ atoms, are moved in one direction, while the σ^+ traps ($m_S = -\frac{1}{2}$ atoms) are moved in the other direction. If this process is applied to an optical lattice filled with atoms, the consequence is that atoms moving in one direction will collide with those moving in the other direction. The optical lattices can then be moved back to their original positions, and the atoms will end up where they started, except that those atoms which have collided will have picked up an additional phase.

For example, consider the simple case of two atoms in neighboring traps, and suppose that the lattices are adjusted in such a way that atoms in state $|0\rangle$ move right, while those in state $|1\rangle$ move left. If the atoms are numbered from left to right, then a collision will only occur if the first atom moves right (and so is in state $|0\rangle$) and the second atom moves left (and so is in $|1\rangle$). Thus the overall evolution (neglecting global phases and any background evolution due to the ordinary energy difference between $|0\rangle$ and $|1\rangle$) is

$$|00\rangle \rightarrow |00\rangle \qquad |01\rangle \rightarrow e^{-i\phi}|01\rangle \qquad |10\rangle \rightarrow |10\rangle \qquad |11\rangle \rightarrow |11\rangle, \qquad (10.2)$$

which is a phase gate

$$U_\phi = \begin{pmatrix} 1 & 0 & 0 & 0 \\ 0 & e^{-i\phi} & 0 & 0 \\ 0 & 0 & 1 & 0 \\ 0 & 0 & 0 & 1 \end{pmatrix}. \qquad (10.3)$$

Choosing $\phi = \pi$ gives a gate very similar to the controlled-Z gate, and it is not surprising that this gate is a universal two-qubit logic gate.

It is important to note that there are several possible ambiguities in the definition of phase gates, both within the optical lattice literature, and also comparing this with other techniques. Firstly there is some variation as to whether there is a plus or minus sign in front of the phase shift, and secondly there is very considerable variation as to *which* state the additional phase is applied to. Finally, some treatments explicitly include the ordinary background evolution. None of this is ultimately very important, as all these definitions are related by simple single-qubit gates. Nevertheless, it is necessary to keep a careful eye out. It is also important to recall that collision gates are coherent processes, and so the same description can be applied to atoms in superposition states, as we will explore next.

10.8 Massive entanglement

In the discussion above we only considered a pair of atoms. It is easy to see that entanglement can be generated in such a system. For example

$$
\begin{aligned}
|00\rangle \xrightarrow{\;H^{\otimes 2}\;} |+\rangle|+\rangle &= (|00\rangle + |01\rangle + |10\rangle + |11\rangle)/2 \\
\xrightarrow{\;U_\pi\;} &\ (|00\rangle - |01\rangle + |10\rangle + |11\rangle)/2 \\
&= (|0\rangle|-\rangle + |1\rangle|+\rangle)/\sqrt{2} \\
&= (|+\rangle|0\rangle - |-\rangle|1\rangle)/\sqrt{2},
\end{aligned}
\tag{10.4}
$$

which is a maximally entangled Bell state, even if it is written in a slightly unusual basis. It could be converted to a more conventional Bell state by applying a selective Hadamard gate to either of the two qubits, but this is difficult in an optical lattice as the two atoms are very close together.

The situation becomes even more interesting in lattices containing very large numbers of atoms. Since both the single-qubit gates and the two-qubit entangling gates are applied to *all* the qubits in parallel, this provides a simple route to extremely large entangled states, known as *cluster states*. For three atoms the phase gate looks like

$$
U_\pi =
\begin{pmatrix}
1 & 0 & 0 & 0 & 0 & 0 & 0 & 0 \\
0 & -1 & 0 & 0 & 0 & 0 & 0 & 0 \\
0 & 0 & -1 & 0 & 0 & 0 & 0 & 0 \\
0 & 0 & 0 & -1 & 0 & 0 & 0 & 0 \\
0 & 0 & 0 & 0 & 1 & 0 & 0 & 0 \\
0 & 0 & 0 & 0 & 0 & -1 & 0 & 0 \\
0 & 0 & 0 & 0 & 0 & 0 & 1 & 0 \\
0 & 0 & 0 & 0 & 0 & 0 & 0 & 1
\end{pmatrix}
\tag{10.5}
$$

and the entangling transformation performs

$$
|000\rangle \xrightarrow{\;H^{\otimes 3}\;U_\pi\;} \frac{|000\rangle - |001\rangle - |010\rangle - |011\rangle + |100\rangle - |101\rangle + |110\rangle + |111\rangle}{2\sqrt{2}}
$$
$$
= \frac{|+\rangle|0\rangle|-\rangle - |-\rangle|1\rangle|+\rangle}{\sqrt{2}},
\tag{10.6}
$$

which is an example of a class of three-qubit entangled states called Greenberger–Horne–Zeilinger states, or GHZ states. With larger numbers of qubits the resulting states are very complex and are called cluster states, and a more sophisticated way of describing these states is hinted at in the exercises. Even more interesting behavior can occur in multi-dimensional optical lattices, as in this case it is possible to move atoms not just left to right, but also back to front and up and down, permitting much more complex interactions.

The result of these processes is often called massive entanglement, and is interesting for many reasons. Firstly, optical lattices provide one of the simplest routes to extremely large entangled states, which are important when studying the transition between quantum and

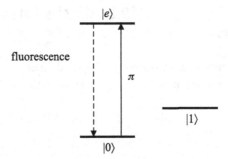

Reading out a qubit state using fluorescence on a cycling transition. The driving field excites qubits in state $|0\rangle$ into an excited state $|e\rangle$, which then fluoresces back to $|0\rangle$. Systems in state $|1\rangle$ are out of resonance with the driving field and are not affected.

classical physics. Secondly, the detailed form of the entangled states produced turns out to be surprisingly useful, with obvious applications in three areas of quantum information processing, namely error correction, quantum simulation, and the implementation of so-called measurement-based quantum computers. The details of these topics are, however, beyond the scope of this book.

10.9 Readout

The final stage which must be considered in any proposed implementation of a quantum computer is, of course, readout: there is no point in running a quantum computation if the final result cannot actually be obtained. This final stage is relatively simple with ion traps, as it is possible to measure the quantum states of ions with accuracy and selectivity. The basic idea is to detect the fluorescence from the ion using optical transitions.

This might seem impossible, as the qubit is deliberately implemented using two levels which do not fluoresce: if they did then they would swiftly relax and the quantum information would be destroyed. The solution is to use a laser to drive the atom from one of the qubit states, say $|0\rangle$, to a third level, which does fluoresce strongly. The simplest situation occurs when the third level decays directly back to $|0\rangle$, an example of a cycling transition, as shown in Figure 10.5. In this case strong fluorescence will be seen at the driving frequency if and only if the qubit is in state $|0\rangle$; as this fluorescence is emitted in random directions it is easily distinguished from the driving laser field. For a superposition state

$$|\psi\rangle = \alpha|0\rangle + \beta|1\rangle \tag{10.7}$$

this process acts as a projective measurement: the system shows fluorescence, and ends up in state $|0\rangle$ with probability $|\alpha|^2$, while no fluorescence is seen, and the atom ends up in state $|1\rangle$, with probability $|\beta|^2$. (The fact that the non-occurrence of fluorescence acts to project the qubit into state $|1\rangle$ is sometimes called a *null measurement*.)

It is, of course, possible to use very similar methods with trapped atoms, but with optical lattices the characteristic problem of distinguishing between different atoms remains. Individual ions in a trap can be distinguished by their positions using a simple microscope, but atoms in an optical lattice are much closer together. It is also not simple to focus the driving laser on a single atom in a lattice. However, these problems can in principle be overcome, either by using sub-diffraction imaging methods,[2] or by using the same selective excitation techniques described in Section 10.5.

Further reading

Atom and ion traps are extensively described in Foot (2005). A physical model of the rotating Paul trap is described in Rueckner *et al.* (1995), while a historical review of optical trapping can be found in Ashkin (1997). Microtraps are described in Hughes *et al.* (2011).

Trapped ions have been considered as potential implementations of quantum computers for many years, beginning with the invention of the two-qubit gate (Cirac and Zoller, 1995) described above; an introductory summary can be found in Ozeri (2011). Implementations using ^{40}Ca^{+} ions, discussed briefly in Part I, as well as the closely related ^{43}Ca^{+} ion, are described in Blatt *et al.* (2004), and detailed reviews of recent work with trapped ions can be found Häffner *et al.* (2008) and Wineland and Leibfried (2011).

There has also been a great deal of research into quantum computing with atoms trapped in optical lattices (Jessen *et al.*, 2004) following the invention of the collision gate (Jaksch *et al.*, 1999). For many years progress was limited by difficulties in resolving individual atoms at such close spacings, but recent breakthroughs include site-resolved readout (Bakr *et al.*, 2010; Sherson *et al.*, 2010) and excitation (Weitenberg *et al.*, 2011).

Exercises

10.1 Write down the potential energy function for a group of ions (each of mass M and charge $+e$) in a linear Paul trap, with strong radial and weak axial confinement. You may assume that the trap potentials are harmonic.

10.2 What effect does the motion of an ion have on its spectral lines if the ion is traveling in free space? What changes if the ion is confined in a harmonic trap?

10.3 Draw a quantum network based on the collisional phase gate U_ϕ, equation (10.3), to implement a controlled-NOT gate with the first qubit as control and the second qubit as target.

---

[2] Although it is frequently assumed that diffraction provides a fundamental resolution limit, it is actually fairly simple to go beyond this if a sufficiently large number of photons are detected.

10.4 Show that the collisional phase gate U_π can be written as $|0\rangle\langle0| \otimes Z + |1\rangle\langle1| \otimes \mathbb{1}$. Hence show that the "massive entanglement" state of a system of two atoms can be written as

$$(|0\rangle Z + |1\rangle)(|0\rangle + |1\rangle),$$

neglecting normalization. How would you write the equivalent state for three atoms? Multiply this out to show that you agree with equation (10.6).

11 Nuclear magnetic resonance

Liquid state nuclear magnetic resonance (NMR) is little studied in most physics courses, but has recently become of considerable interest as a method for building small quantum computers. The underlying ideas were introduced in Part I of this book, but these were limited to systems with a single nuclear spin, and thus a single qubit. Here we expand these ideas to cover systems with two or more qubits.

11.1 Qubits

The basic idea behind NMR quantum computing is that the two spin states of a spin-$\frac{1}{2}$ nucleus provide a natural implementation of a qubit: indeed, it is such a natural implementation that a qubit is sometimes referred to as a spin. To obtain more than one qubit, just use more than one spin; as discussed below this in practice means choosing a suitable molecular system with the desired combination of nuclear spins. There are, however, flaws in this naive approach, which can be traced back to the low frequencies (and thus low energies) of NMR transitions,

$$\Delta E = h\nu = \hbar\gamma B, \tag{11.1}$$

where γ is a constant characteristic of the nucleus called the *gyromagnetic ratio*. This energy gap arises from the Zeeman interaction of the nuclear spin with an externally applied magnetic field, and for reasonably accessible field strengths (up to around 20 T) lies in the range up to 1 GHz.

While these low frequencies make controlling the radiation very easy, the corresponding long wavelengths mean that it is impossible to directly distinguish between different nuclei according to their positions in space. Instead, it is necessary to use the different transition frequencies observed for different nuclei for qubit selection. This is easy as long as different qubits are represented by different nuclear species, as these have different gyromagnetic ratios, leading to very different frequencies. However, there are only a small number of distinct spin-$\frac{1}{2}$ nuclei available, and a computer of any reasonable size will have to represent several different qubits with the same nuclear type.

It is in principle possible to use the methods of magnetic resonance imaging (MRI) to make the transition frequency depend on the spatial position of a nucleus, and this approach has been considered. However, the spatial resolution required goes far beyond that normally achieved, and this approach seems very difficult in practice. Fortunately the exact field experienced by a nucleus is not simply equal to the externally applied field, but also depends on local fields. Roughly speaking, the external field induces a current in the electrons surrounding a nucleus; this current produces an additional field which acts to shield the nucleus from the applied field. These shielding effects, and thus the value of the transition frequency, will depend on the details of the nuclear environment, an effect known as the *chemical shift*.

A second problem of the low frequencies is that the energy of the corresponding photons (a few microelectronvolts) is so low that it is not currently practical to detect a single radio frequency photon. Thus we cannot detect a single nuclear spin, and instead have to use an ensemble of identical independent nuclei. This greatly limits the range of systems available to us, and as we will see has very significant consequences for both initialization and readout.

The solution adopted in most NMR quantum computing studies to date is to use fairly dilute solutions of small molecules in inert solvents. (There have been some studies involving solid state systems or more complex systems such as liquid crystals, but these are not considered further here.) Each molecule will contain a number of different spin-$\frac{1}{2}$ nuclei, each of which can be used as a qubit. (Spin-0 nuclei can be ignored for obvious reasons; nuclei with spins greater than $\frac{1}{2}$, usually called *high-spin* nuclei, can also be largely ignored for more subtle reasons which we neglect here.) As every molecule is chemically identical, there are a very large number of copies of the quantum computer, and the experiment controls these in parallel. It might seem that different molecules would in fact be subtly different (due to effects such as internal motions and the orientation of the molecules with respect to one another and the applied field), but all these effects are averaged out by the rapid molecular tumbling which occurs in solution. Furthermore, it turns out that the interactions between spins in different molecules are averaged out by the same tumbling process, and so molecules in liquids provide an ensemble of identical independent copies, as required.

Example 11.1 Two spin systems used in the early days of NMR quantum computing are typical of the sort of molecules used to implement two-qubit systems. The first example, ^{13}C-labeled chloroform, contains a single ^{1}H nucleus and a single ^{13}C nucleus, both of which are spin-$\frac{1}{2}$ (in unlabeled chloroform the carbon nucleus would normally be the dominant ^{12}C isotope, which is spin-0). The molecule also includes three chlorine nuclei, which will be a mixture of ^{35}Cl and ^{37}Cl; all of these chlorine nuclei are spin-$\frac{3}{2}$ and can be ignored. A variety of solvents have been used; usually such solvents are *deuterated*, replacing any ^{1}H nuclei with the spin-1 nucleus ^{2}H. As the two spin-$\frac{1}{2}$ nuclei used to implement qubits are of different nuclear species this is called a *heteronuclear* spin system.

By contrast the second example, partially deuterated cytosine, contains two ^{1}H nuclei, and so is called a *homonuclear* spin system; as usual these two nuclei have different chemical shifts, and so different resonance frequencies. Like chloroform, cytosine actually contains a host of other nuclei, but these are all spin-0 (^{12}C and ^{16}O) or spin-1 (^{14}N and ^{2}H). The required partial deuteration is easily achieved by dissolving the cytosine in deuterated water, replacing the three labile ^{1}H atoms bound to nitrogen atoms with ^{2}H.

11.2 Initialization

The obvious method to initialize a spin system is simply to cool the spins directly into their thermodynamic ground state (since a spin-$\frac{1}{2}$ particle is a natural qubit there is no question about cooling into one specific sublevel). However, a comparison of the transition energy (say 1 μeV) with kT at room temperature (around 25 meV) reveals that this approach will require temperatures around 1 mK. This temperature is perfectly attainable, but clearly not for solutions of small molecules.

There are three potential ways around this problem. The first is to switch to NMR studies of solid state systems, where direct cooling may be practical. The second is to find some cunning method to obtain a non-equilibrium population of the spin states. The third is, quite simply, to cheat! With ensemble systems it is possible to use so-called *pseudo-pure* states, which appear to behave like pure states. All three approaches have been used, but the method of pseudo-pure states is by far the most common.

The method of pseudo-pure states relies on two basic ideas to generate a state which, at least superficially, behaves like a pure state. The first is that it is possible to use a combination of unitary and non-unitary operations to convert the thermodynamic equilibrium state of a spin system into a state where all the different energy levels have the same population, except for the ground level of the whole system which has a slightly higher population. The second key fact is that the NMR experiment is not sensitive to spin systems where all the levels have the same population, so that the only signal actually observed from this state arises from the small excess population in the ground level. Thus the large ensemble of spins in a highly mixed state gives the same result as a much smaller ensemble of spins in a pure state.

To take a concrete example, consider a molecule containing two spin-$\frac{1}{2}$ nuclei, each of the same nuclear species and with small interactions such as the chemical shift ignored. The Zeeman interaction means that the ground level $|00\rangle$ will lie at an energy $h\nu$ below the unsplit position, while the level $|11\rangle$ will lie at $h\nu$ above the unsplit position. The two levels $|01\rangle$ and $|10\rangle$ will be degenerate at the original energy. Clearly the state $|00\rangle$ will have the largest population at thermal equilibrium, but the population differences will be small. In the high-temperature limit $kT \gg h\nu$, which applies in liquid state NMR, the fractional populations can be written as $1/4 + \epsilon$, $1/4$, $1/4$ and $1/4 - \epsilon$, with $\epsilon \sim 10^{-5}$, and the resulting density matrix of the system is

$$\rho_B = \sum_j P_j |j\rangle\langle j| = \begin{pmatrix} 1/4 + \epsilon & 0 & 0 & 0 \\ 0 & 1/4 & 0 & 0 \\ 0 & 0 & 1/4 & 0 \\ 0 & 0 & 0 & 1/4 - \epsilon \end{pmatrix} \tag{11.2}$$

while the desired pure state is

$$\rho_{00} = |00\rangle\langle 00| = \begin{pmatrix} 1 & 0 & 0 & 0 \\ 0 & 0 & 0 & 0 \\ 0 & 0 & 0 & 0 \\ 0 & 0 & 0 & 0 \end{pmatrix}. \tag{11.3}$$

The method of pseudo-pure states works by averaging the populations of all the levels except $|00\rangle$. The resulting state is

$$
\rho_{PP} = \begin{pmatrix} 1/4 + \epsilon & 0 & 0 & 0 \\ 0 & 1/4 - \epsilon/3 & 0 & 0 \\ 0 & 0 & 1/4 - \epsilon/3 & 0 \\ 0 & 0 & 0 & 1/4 - \epsilon/3 \end{pmatrix} = (1/4 - \epsilon/3) \times \mathbb{1} + (4\epsilon/3) \times \rho_{00}
$$

(11.4)

and since the maximally mixed term is not observable, this state looks just like a pure state, except that the signal is only $4\epsilon/3$ times as large as the signal from a true pure state.

The pseudo-pure state approach works well with small spin systems, but has major problems when applied to larger spin systems. This point will be discussed in more detail later.

11.3 Decoherence

At first glance it seems that decoherence should not be a major problem for NMR quantum computing. The ultimate limit is provided by the spontaneous emission lifetimes of the spin states, and these are around 10^9 years. Real systems are not quite so extreme, as NMR relaxation is dominated by stimulated emission and absorption, arising from random fluctuations in a spin's environment, but relaxation times for spin-$\frac{1}{2}$ nuclei are still very long, typically around 1 s.

This coherence time is far greater than that observed in many quantum systems, but this is not as important as it might seem. While the long coherence time makes experiments relatively simple, what matters for quantum computation is not the coherence time itself, but rather the ratio of the coherence time to the gate time. This ratio determines the number of gates which can be performed before the system has decohered away, and NMR gates are also very slow in comparison with those in systems of trapped atoms or ions.

A more subtle advantage of NMR is that the use of ensemble systems means that decoherence appears differently in NMR quantum computers than in other systems. In a conventional quantum computer, decoherence introduces the possibility that the computer will make some random transition, and so return the wrong answer at the end of the computation. In ensemble computers, however, different members of the ensemble can return different answers, with the observed result being an average over the whole ensemble. With luck the wrong answers produced will largely cancel out, so that decoherence appears simply as a reduction in the signal strength rather than as an actual wrong answer. Real life is rarely quite so kind, but there is some truth in this idea.

11.4 Universal logic

As usual, the problem of implementing universal logic comes down to the problem of implementing single-qubit and two-qubit gates. Single-qubit gates can be implemented using resonant radio frequency pulses, and this topic was considered in detail in Part I of this book. The only subtle points arise from the problem of applying these gates to individual qubits, rather than the whole spin system, and here we simply assume that sufficient frequency dispersion exists to allow frequency selection to be achieved.

Two-qubit gates are much more interesting, as these require some sort of spin–spin interaction. The obvious source is the direct interaction between the magnetic dipoles of two nuclei, but this interaction depends on the angle between the internuclear vector and the magnetic field, and is completely averaged out by rapid molecular tumbling (this is in essence the reason why different molecules can be treated as being independent of one another). Instead of this, the key interaction in liquid state NMR studies is *scalar coupling*, also known as J-coupling. This is closely related to the electron–nuclear hyperfine interaction, and is mediated between different nuclei by shared valence electrons within the molecule. The key result is that the J-coupling interaction does survive molecular tumbling, and provides a coupling between spins in the same molecule. However, because it is a correction to the direct coupling term, it is typically rather small (rarely more than 500 Hz).

The scalar coupling naturally has the *Heisenberg* form expected for an exchange interaction,

$$\boldsymbol{\sigma}_1 \cdot \boldsymbol{\sigma}_2 = \sigma_{1x}\sigma_{2x} + \sigma_{1y}\sigma_{2y} + \sigma_{1z}\sigma_{2z}, \tag{11.5}$$

but in most situations is *truncated* to the *Ising* form $\sigma_{1z}\sigma_{2z}$ by the larger Zeeman interactions. (Strictly speaking the terms Heisenberg and Ising should only be used for extended networks of couplings of these forms, and not applied in systems with only a few spins, but this slight abuse of language is common in NMR texts.) This is nothing more than first-order perturbation theory, where a small perturbation is seen to have little or no effect on the eigenstates of a system, while slightly shifting the eigenvalues (energy levels), so that the Hamiltonian becomes effectively diagonal. Thus the overall Hamiltonian of a two-spin system can be written as

$$\mathcal{H}/\hbar = \omega_1 \frac{\sigma_{1z}}{2} + \omega_2 \frac{\sigma_{2z}}{2} + \omega_{12}\frac{\sigma_{1z}\sigma_{2z}}{4}, \tag{11.6}$$

where energies have been written in angular frequency units as usual, and the factors of $\frac{1}{2}$ are necessary as $\hbar\omega$ is the energy gap between the $\pm\frac{1}{2}$ states, not the shift of the individual states. Note that many treatments of NMR QC use traditional NMR conventions, where the $\hbar$ is simply dropped and Pauli matrices are replaced by other operators which are equivalent up to a constant factor, so the exact form of the Hamiltonian can seem quite variable.

Example 11.2 Using the so-called *product operator* notation which is standard in conventional NMR texts the Hamiltonian above would be written as

$$\mathcal{H} = 2\pi \nu_I I_z + 2\pi \nu_S S_z + \pi J\, 2I_z S_z, \tag{11.7}$$

where the factor of $\hbar$ has been dropped, the elementary operators are defined by

$$I_z = \tfrac{1}{2}\sigma_{1z} \qquad S_z = \tfrac{1}{2}\sigma_{2z}, \tag{11.8}$$

and energies are labeled in frequency units, not angular frequencies. The use of I and S to label the two spins is traditional. The origin of this convention is not entirely clear, but probably stems from earlier treatments of ENDOR (Electron Nuclear DOuble Resonance) experiments, in which the nuclear spin was labeled as I and the electron spin as S; it was subsequently popularized in the works of the aptly initialled Ionel Solomon. The notation for the third and subsequent spins is very variable, but the letters I, R, S and T are frequently used. The placement of factors of two in product operator notation is unusual at first sight, but makes slightly more sense in its full context.

This Hamiltonian contains a spin–spin interaction, and so in principle will permit universal quantum computation, but the form is rather more complicated than one might wish, and it would be nice to be able to "sculpt" it into a more desirable form. This can be achieved with spin echoes, which were discussed in detail for single-qubit systems at the start of this book. The basic idea is to allow the system to evolve under a Hamiltonian for a time $\tau/2$, apply a NOT gate, allow the system to evolve for another time $\tau/2$, and then apply a final NOT gate (this final NOT gate is not always included, but it makes the overall analysis slightly simpler). In a single-qubit system this will refocus the evolution under the Zeeman splitting (rotation around the z axis at the Larmor frequency). A simple way to think about this is that by interconverting $|0\rangle$ and $|1\rangle$ the NOT gate causes the spin to precess *backward* during the second time period $\tau/2$.

In a two-qubit system several different spin echoes are possible, as NOT gates can be applied to either or both of the two spins. The effect on the Zeeman terms is obvious, and each of these will be refocused if a pair of NOT gates is applied to the corresponding spin. It only remains to consider what happens to the spin–spin coupling. If a pair of NOT gates is applied to one of the two spins, then the coupling will also be refocused in the same way. If, however, NOT gates are applied to both spins then the coupling is unaffected by the NOT gates, and so its evolution adds up during the whole spin echo. In effect, the NOT gates applied to one spin reverse the coupling evolution during the second $\tau/2$ period, but the NOT gates applied to the other spin reverse it again, leaving it overall unaffected. Thus the effect of the spin echo is equivalent to evolving under the *average Hamiltonian*

$$\mathcal{H}_{av}/\hbar = \omega_{12}\frac{\sigma_{1z}\sigma_{2z}}{4} \tag{11.9}$$

for the whole time period τ.

Example 11.3 The networks below show how to isolate any single term in a two-spin system as the sole surviving term in the NMR Hamiltonian:

$$\tfrac{1}{2}\omega_1\sigma_{1z} \tag{11.10}$$

$$\tfrac{1}{2}\omega_2\sigma_{2z} \tag{11.11}$$

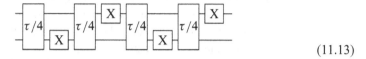

$$\frac{1}{4}\omega_{12}\sigma_{1z}\sigma_{2z} \tag{11.12}$$

To refocus *all* the terms it is necessary to ensure that the NOT gates do not coincide, dividing the time τ up into four equal periods. One possible network is shown below:

$$\tag{11.13}$$

More complex patterns on NOT gates can be used to scale down individual terms rather than eliminating them completely.

In a multi-spin system there will in principle be couplings between all the pairs of nuclei, and so the Hamiltonian takes the general form

$$\mathcal{H}/\hbar = \sum_{j} \omega_j \frac{\sigma_{jz}}{2} + \sum_{j<k} \omega_{jk} \frac{\sigma_{jz}\sigma_{kz}}{4}. \tag{11.14}$$

Using more complex patterns of spin echoes it is possible to sculpt the Hamiltonian almost at will, refocusing undesired terms while keeping those that are wanted.

It might seem that a NMR quantum computer would require every spin to be directly coupled to every other spin, but this is in fact not the case. It is, however, necessary that every spin be directly or indirectly coupled to every other spin, where indirect coupling means that two spins are connected to one another through a chain of directly coupled intermediate spins. For example, a linear chain of spins, each of which is coupled to its immediate neighbors, is sufficient. To see that this is true, note that a (logical) qubit can be moved around the (physical) spin system using quantum SWAP gates, so that any two qubits can be made to interact even if there is no direct coupling between the spins they are currently located on.

11.5 Readout

The basic readout mechanism used in NMR quantum computing is to acquire an NMR spectrum. As we will see, this ensemble process is quite different from the conventional projective measurements that are normally discussed in quantum mechanics, and this has significant consequences.

We begin by considering the conventional NMR spectrum acquired from a spin system in its thermal equilibrium state. A single isolated spin will give a signal at its Larmor frequency. In elementary treatments of NMR this signal is sometimes discussed as if it were a conventional absorption signal, arising from the absorption of photons resonant with the NMR transition. In fact, the NMR signal is a coherent response to the applied radiation field, and this naive approach is entirely inappropriate. Despite this fact the language of "absorption" (by spin systems in the ground state) and "emission" (by excited spin states) is widely used.

A slightly better picture of NMR signal detection is provided by the vector model, introduced in Section 3.4. An ensemble of isolated spins can be treated as a classical magnetic moment, which behaves like a macroscopic version of the Bloch vector. In particular, the Bloch vector corresponding to a superposition state precesses around the z axis at the Larmor frequency, and the magnetic moment will precess around the applied magnetic field in the same way. This precessing magnetic moment will induce an oscillating electric field in the NMR detection coil, and this field is detected as the NMR signal.

Note that NMR signals can only be seen from superposition states, and spin systems in eigenstates cannot be detected. However, applying a 90° pulse to a spin system in an eigenstate produces a detectable superposition. The phase of this superposition will be different for spins starting in states $|0\rangle$ and $|1\rangle$, and this can be detected as the phase of the oscillating signal. As usual the absolute phase of the signal has no meaning, but the relative phase of the excitation pulse and the detected signal can be measured, allowing the states $|0\rangle$ and $|1\rangle$ to be distinguished. (In real experiments even this relative phase can only be measured up to an unknown but constant offset; this offset can, however, be measured using a spin system in a known initial state, and then corrected in subsequent experiments.)

The NMR sample is a macroscopic ensemble of molecules, which at thermal equilibrium can be considered as a mixture of the states $|0\rangle$ and $|1\rangle$. After applying a 90° rotation the coherent signals from this mixture will mostly cancel out, and the component of the mixture corresponding to the maximally mixed state will give no overall signal. There will, however, be a small excess in the low-energy state, and so a small signal corresponding to this excess population will be seen. For this reason the signal seen from a state at thermal equilibrium is referred to as an "absorption" spectrum.

A pair of isolated spins with different chemical shifts will give rise to two separate signals, one for each spin. If these spins form a homonuclear spin system then these two lines will be visible in the same spectrum, but if they belong to different nuclear species (a heteronuclear spin system) it will not normally be possible to observe them both in the same experiment. The size of the signals will depend on the number of spins in the ensemble, the polarization of the spins (that is, the size of the excess population in each spin's ground state), and the intrinsic sensitivity of the NMR apparatus. In general the relative size of two NMR signals in a homonuclear spectrum can be compared directly, but the relative sizes of signals from two different nuclear species can only be compared after careful calibration.

In a system of two coupled spins the spectrum will be slightly more complex, as the transition frequency of each spin now depends on the state of the other spin. At thermal equilibrium the two spin states of the other spin will be almost equally populated, and so the signal is split into two lines of equal intensity, called a doublet, with a splitting equal to the J-coupling constant ω_{12}. A stylized spectrum for a homonuclear system is shown in Figure 11.1. In a system with more than two coupled spins, each line will be divided into a group of lines called a multiplet.

The situation is very similar when NMR spectra are used for readout of the final state of an NMR quantum computer, except (of course) that the sample does not begin at thermal

Fig. 11.1 Stylized NMR spectrum (that is, signal intensity as a function of frequency) from a homonuclear two-spin system: the two groups of signals correspond to transitions of each spin, while the splitting indicates the dependence of the transition frequency on the state of the other spin. As spectra are measured on ensembles, every possible transition is seen; for thermal states at high temperatures all transitions have essentially the same intensity. The signal intensity is measured in arbitrary units, and for this schematic spectrum no explicit frequency axis is shown. Note that conventional NMR experiments use several idiosyncratic conventions; in particular, NMR spectra are plotted such that the "absorption" spectrum which occurs at thermal equilibrium is shown as a positive signal, and spectra are frequently plotted with the frequency axis reversed. These conventions are usually also followed in NMR implementations of quantum information processing.

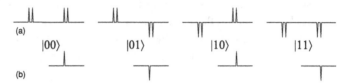

Fig. 11.2 Stylized spectra depicting readout in (a) homonuclear and (b) heteronuclear two-qubit NMR quantum computers for each of the four possible basis states. As before, "absorption" spectra are shown as positive intensities and "emission" spectra as negative.

equilibrium. In this case the state can be determined from the relative intensities of the various lines in the various multiplets, as shown in Figure 11.2 for the case of a two-qubit system. In homonuclear spectra both qubits are excited and observed, and the state of each qubit can be determined from the phase (absorption or emission) of the corresponding multiplet in the spectrum. In heteronuclear spectra only one spin is excited and observed, here assumed to be the second qubit, and the state of the corresponding qubit can be determined directly from the phase of that spectrum. Within each multiplet only one line is observed, and the state of the other qubit can be determined from the identity of the transition corresponding to this line. Both techniques can be generalized to larger spin systems.

When considering NMR readout mechanisms it is vital to remember that NMR experiments are carried out not on a single molecule but on an ensemble of identical molecules. Thus an NMR spectrum does not reveal the state of an individual molecule, but rather the ensemble average over all molecules. An important consequence is that NMR measurements do not cause superposition states to collapse. This immediately explains a common worry about NMR. The description of NMR measurements given above involves observing a rotating magnetization, and measuring its x and y components. Quantum mechanically this corresponds to measuring the expectation values of the x and y components of the spin's angular momentum. As these are non-commuting observables, this ought to be impossible! The solution to this apparent paradox is that the ensemble measurement does not in fact correspond to a projective measurement of a single spin, and so these arguments do not apply.

These ensemble measurements might seem more powerful than conventional projective measurements, but in fact they are usually less useful for two reasons. Firstly projective measurements, followed by classical control based on the result of the measurement, play a key role in schemes such as quantum error correction. More importantly, projective measurements provide an excellent initialization method: just measure a bit, and then flip it if it has the wrong value. If NMR had projective measurements there would be no need to worry about pseudo-pure states.

The lack of projective measurements also causes difficulties for NMR algorithms which end in superposition states rather than basis states; for example the generalization of Grover's algorithm with $k > 1$ satisfying values ends up in an equally weighted superposition of all the satisfying values. While a conventional projective measurement of the system will reveal one of the states in the superposition, with the exact state chosen at random, an NMR measurement will show some sort of average over all the possibilities. In a few cases it is possible to interpret these averages in a straightforward fashion, and in other cases it is possible to use simulations to show that the output spectrum does at least take the expected form. In general, however, this can be a serious problem.

Further reading

NMR quantum information processing is briefly covered in texts such as Bouwmeester *et al.* (2000), Le Bellac (2006), and Stolze and Suter (2008), with particularly detailed coverage in Estève *et al.* (2003). Several introductions have been published in journals (Cory *et al.*, 2000; Jones, 2001) and several reviews have described more recent developments (Vandersypen and Chuang, 2004; Ryan *et al.*, 2008; Suter and Mahesh, 2008; Jones, 2011). Product operators and NMR conventions are well described in standard NMR textbooks, such as Levitt (2008).

Exercises

11.1 Estimate the strength of the magnetic field gradient required to make two ^{1}H nuclei in a molecule (assume a separation of about 0.1 nm) have Larmor frequencies differing by about 100 Hz. Would it be possible to obtain a gradient of this size?

11.2 Show that a Heisenberg coupling in a two-spin system can be approximated by an Ising coupling as long as $|\omega_{12}| \ll |\omega_1 - \omega_2|$.

11.3 Draw an explicit network of gates to implement a controlled-NOT gate in a two-spin system, using only standard single-qubit gates and the gate $U(t)$ which corresponds to free evolution under the system's Hamiltonian for a time t. Draw an implementation of a NOT gate that takes the same length of time.

11.4 Consider a system of three coupled spins. Write down the Hamiltonian and then design a spin echo sequence such that the average Hamiltonian is reduced to a single coupling term between the second and third spins.

11.5 Return to a one-spin system, and design a spin echo-style sequence that will reduce the spin's apparent Larmor frequency to one half of its true value. Is it possible to change the sign of a spin's apparent Larmor frequency? What are the limits on the possible range of scalings? Can coupling strengths be rescaled in the same way?

12 Large-scale quantum computers

Trapped ions, trapped atoms, and NMR spin systems are all fine ways of building small "toy" quantum computers, each with its own advantages and disadvantages. There are also many other techniques which have been suggested, although these three so far remain in the lead for general-purpose quantum computing. However, the most powerful general-purpose quantum computers constructed to date have only about a dozen qubits, and this is not nearly large enough to make quantum computers useful rather than merely interesting. (Larger systems have been used to demonstrate particular quantum information processing techniques, but these cannot as yet be used to implement arbitrary quantum algorithms.)

Although it is not completely clear how complex a general-purpose quantum computer needs to be, it is clear that such a device will involve thousands or even millions of qubits, rather than the dozens involved today. It is, therefore, important to consider whether there is any hope of scaling up these technologies to useful sizes, and we will consider each of the three approaches in turn, before turning briefly to alternative technologies which have not been discussed so far.

12.1 Trapped ions

Trapped ions initially look very promising as a candidate for scaling up, as it is possible to trap thousands of ions while keeping a reasonable distance between them. Early experiments relied on particular tricks which only work with systems of two ions, but this is not true of more recent work, and there is no reason in principle why these large strings of ions could not be controlled. One major issue, however, turns out to be the question of implementing logic gates in parallel in different parts of the computer. While this ability is not required in an ideal world, it is essential in large-scale devices which rely on error correction.

For single-qubit gates this means that each ion should be controlled by its own laser beam, rather than directing a single laser beam to different regions of the apparatus. The idea of a system with a few thousand lasers may sound ridiculous, but in fact is surprisingly reasonable, as they can all be generated from a single master source and controlled by small mirrors. Such devices, called micromirror arrays, are well developed for more conventional applications, and development has already begun on systems with hundreds of thousands of controllable mirrors.

Two-qubit gates are, however, more tricky, as the data bus provided by the common vibrational modes of the ions proves to be a limit as well as an advantage. In the standard scheme this bus can only hold a single qubit, and so conventional ion traps can only implement a single two-qubit gate at a time. To get round this limit it is necessary to move to a much more complex design, involving very large numbers of ion traps each holding a small number of ions. These traps can then communicate by moving individual ions between them. Moving the ions inevitably heats them up slightly, but this can be corrected by using *sympathetic cooling*: the ions used to implement qubits are interspersed with ions of a different atomic species, and laser cooling of these neighbor ions will cool the whole ion string without affecting the internal states of the qubits. Remarkably it has already proved possible to demonstrate that ions can be moved between traps while retaining quantum coherence, but while this is a significant first step it is only a first step.

More recently a different approach to implementing quantum logic gates has been demonstrated. This uses microwave fields to drive transitions directly, with spatial localization achieved by generating the microwaves using the trap electrodes, so that the interaction occurs in the near-field limit where conventional resolution limits do not apply. The rapid progress being made makes trapped ions a fairly plausible technology for the near future, but the difficulties remaining are substantial.

12.2   Optical lattices

Until recently, the problem with optical lattice implementations was not so much scaling them up as getting them to work at all! Recent progress has, however, been remarkable, although as with ion traps substantial difficulties remain.

The lack of selective gates initially appeared to be a critical problem, but this has now been solved for single-qubit gates. The two-qubit collision gate at first sight appears to be intrinsically parallel, but if selective singlet-qubit gates are available then spin-echo techniques can be used in much the same way they are used in NMR to sculpt this parallel gate into a more useful form. The other extreme is to abandon optical lattices and use individual atom traps, which can be built to some desired scale, and which permit the direct addressing of individual atoms, but this approach throws away the massive intrinsic parallelism which is a key idea of optical lattices.

It should be noted that it may not be necessary to be able to perform all gates selectively. Perhaps the most interesting ideas are based on the method of one-way computation using cluster states. In this model, parallel gates are used to generate massive entanglement, and then selective measurements are used to implement the computation. This model only requires atom-selective measurements, and these are perhaps less challenging than atom-selective gates.

12.3 NMR

NMR remains by far the simplest way to implement small quantum computations,[1] but there are formidable problems in scaling it up. The most obvious problem is that of initialization: while the pseudo-pure state method works well with small systems, it is ultimately a cheat and cannot be scaled up to large systems. The fraction of the ensemble found in the "excess" population of ground state falls off as the number of possible states increases, and so the signal strength falls off exponentially with the number of qubits involved. With more than a few dozen qubits this exponential fall-off renders the approach completely hopeless.

One way around this is to switch to solid state NMR, where the sample can be cooled sufficiently. Experiments on this approach have begun, but the problem is very challenging as the move to the solid state replaces the simple Hamiltonian found in liquid state NMR systems with a much more complex form. A more subtle approach is to use non-equilibrium spin states in liquid state samples, and recent experiments have shown that it is possible to generate a two-qubit NMR quantum computer in a pure ground state; however, this approach is currently limited to two-qubit systems.

Another problem is the lack of projective measurements. As previously noted, this has serious problems for effective error correction, which seems to require this step. In fact this is not true, as projective measurements and classical control can be replaced by quantum control and a qubit reset mechanism. However, all current NMR initialization methods can only be applied at the start of a computation, and resetting a qubit in the middle of a computation is not as yet possible. There are some very speculative ideas for achieving single-qubit readout using spin-sensitive atomic force microscopes, but little has been demonstrated so far.

12.4 Other approaches

There are many other technologies which have been suggested as a means for implementing quantum computation. This reflects the fact that almost any physical interaction between qubits provides a universal two-qubit gate, and so any system with sufficient control of decoherence to implement single qubits could in principle be used for quantum computation. In practice, however, many of these proposals are quite unrealistic, either because the decoherence times are simply too short to permit true coherent control of qubits, or because the method used to bring two qubits together also acts to increase the decoherence rate, so that the qubit implementation is no longer viable. Still other techniques provide plausible

---

[1] Implementing even the simplest quantum computation with an ion trap is a very challenging task; by contrast, any competent NMR spectroscopist should be able to get Deutsch's algorithm working on a conventional NMR spectrometer in less than a week.

implementations of one- and two-qubit devices, but use techniques which do not scale up to larger numbers of qubits.

When assessing these proposals it is important to remember that the conventional circuit model of quantum computation, which has been used throughout this text, is not the only possible approach. One-way quantum computation, also known as measurement-based quantum computation, has been briefly mentioned above: this uses a simple set of one- and two-qubit gates to prepare a specific highly entangled state, which is then manipulated solely by measuring individual qubits, with the choice of measurement to be made depending on both the algorithm being implemented and the outcome of previous measurements. A related approach, teleportation-based quantum computation, uses "entangling measurements," which will be discussed briefly in Part III, to perform the desired manipulations. Remarkably these two schemes can be used even in systems with quite high probabilities of error, as long as the occurrence of the error is detectable, providing a sort of intrinsic fault tolerance.

A third approach, requiring even less control of the quantum system, is provided by the ideas of quantum cellular automata; these permit quantum computations to be performed in systems where the great majority of the qubits are not individually addressable, but can only be manipulated en masse. Finally, adiabatic quantum computation uses an entirely different approach in which the Hamiltonian describing a quantum system is slowly swept from some initial background form to some other sculpted form, simulating a physical system of interest; the state of the system remains in the ground state of the instantaneous Hamiltonian as long as the sweep is adiabatic, permitting the ground states of arbitrary Hamiltonians to be efficiently located.

This enormous breadth, both of models of computation and of potential technologies, makes it very difficult to summarize the possible approaches, but several themes are clear. One family of approaches uses quantum dots as "artificial atoms"; these have the advantage that their properties can to some extent be designed, but the concomitant disadvantage that it is hard to produce dots in a completely reproducible fashion. A further advantage is that the dots can be placed in desired patterns on surfaces, aiding both discrimination between individual dots and the sculpting of couplings between neighboring dots. Initial experiments have shown some success, but so far have been quite limited by short relaxation times.

A second family of experiments extends the spin physics techniques demonstrated in NMR quantum computing to electron spins; these are more challenging to control coherently because of their larger magnetic moments, but the corresponding larger energy scale makes preparation and readout significantly simpler. With electron spins it is also possible to couple the spin degree of freedom to spatial degrees of freedom, for example using spin-valve transistors, and to modulate spin–spin coupling sizes by changing the shape of the electron's wavefunction. A third family is derived from the coherent manipulation of superconducting quantum interference devices, or SQUIDs. For many years these were largely limited to single-qubit and two-qubit devices, but more recently substantial progress has been made, and a four-qubit general-purpose device has now been demonstrated. Substantially larger SQUID systems have also been used to implement adiabatic computations.

Further reading

A thorough summary of possible technologies for quantum computation is provided by the ARDA Roadmap (Hughes, 2004). There are several collected volumes which describe a range of approaches, such as Estève *et al.* (2003), and several journals have published special editions collecting summary papers (Braunstein and Lo, 2000; Everitt, 2004; Southwell, 2008). A good recent summary comparing all major approaches is Ladd *et al.* (2010). Recent developments in ion traps are described in Ospelkaus *et al.* (2011) and Timoney *et al.* (2011).

PART III

QUANTUM COMMUNICATION

13 Basics of information theory

Scientific progress in physics and mathematics has led to the development of efficient technology for communicating and distributing information in the 20th century. This technology forms one of the cornerstones of information society and our global economy, which are highly reliant on secure methods to transfer and distribute information quickly. To satisfy ever-increasing demands for speed and security, encryption and communication methods are constantly improved and there is an ongoing effort to develop corresponding technology further.

Classical information theory is not usually part of a physics undergraduate degree course. The theory assumes simple classical properties of physical systems and based on those a largely mathematical theory of information is established. The obtained results are independent of the chosen implementation and its physical details. However, one still has to acknowledge that information is physical, that is information carriers, senders, and receivers obey the laws of physics. The notion of quantum information comes about since it turns out that the rules of quantum mechanics violate some of the basic physics assumptions in classical information theory. The consequences of this are many, ranging from improved channel capacities, the possibility of physically secure communication protocols, to invalidating some beliefs about the security of classical communication protocols.

So far the approach followed in this book was to assume idealized physical systems which perfectly implement qubits and then demonstrate how quantum computing algorithms can be realized using perfect qubits. Error correction and fault tolerance were only mentioned briefly. Partly this can be justified by noting that an experimental quantum computation either works or does not, and if it does not work one needs to find and eliminate the error and try again. In quantum communication one can be a bit more pragmatic and allow for an acceptable level of error, which implies that we now must deal with errors more carefully. For instance, decoherence may affect atomic qubits differently than photonic qubits and details of the interaction with an environment leading to decoherence may matter. Also, for the sender and receiver in a quantum communication protocol the action of an eavesdropper is often indistinguishable from decoherence processes. For these reasons the field of quantum information theory is currently still linked with physics. Physical properties of information carriers influence the performance and security of quantum communication protocols qualitatively, and device-independent theories are less well developed than in the classical case. However, as this research field gradually matures, quantum information theory increasingly develops into a field of its own, equally applicable to all physical realizations.

We will now introduce the basics of classical information theory and some of the most important quantum counterparts, pointing out some fundamental differences between them.

This is followed by a chapter on schemes for efficient quantum communication. The violation of basic classical assumptions is then exemplified by showing how entangled states violate Bell's inequalities and thus local realism, one of the cornerstones in classical information theory. Finally, we will discuss physically secure cryptographic communication methods which exploit quantum properties of information carriers. We will explicitly show how indistinguishability of non-orthogonal states of a qubit and two-qubit entangled states can be used to securely transmit information.

13.1 Classical information

Information must be embodied in the state of a physical system and processing of information must be accomplished by dynamical evolution of a physical system. Information is thereby defined by the ability to perform a certain task and quantified by how many *resources* are required to perform a specific *task successfully*. One can, for instance, ask "How many DVDs (the resources storing the information) are needed to store a map of the UK (the task) which specifies the boundaries of each postcode region (success)?" Similarly we can ask questions about quantum information processing, like "Which physical resources permit the transmission of a qubit state $|\Psi\rangle$ from sender Alice to receiver Bob with entanglement fidelity $F = 0.999$?" As can already be seen from these simple examples, numerous different types of resources exist and success can be defined in a number of different ways. It is thus desirable to quantify information in terms which are, to a large extent, independent of the physical realization.

13.1.1 Quantifying classical information

We consider the situation where a sender, usually called Alice, communicates with a receiver, Bob, over a communication channel. We do not wish to make assumptions on how the messages are embodied. We assume the following general setup.

- The sender Alice:
  - She can send one out of N different messages $x_1, \ldots, x_N$ per use of the channel.
  - The probability that Alice chooses message x_j is known and given by p_j. We assume that we do *not know* the physical laws which would allow us to calculate the particular message chosen by Alice.
- The communication channel:
  - The channel is capable of transmitting one of Alice's N messages to the receiver in each use.
  - It can introduce noise and be susceptible to eavesdropping by a third party, Eve.
  - We consider classical and (later) quantum channels.
- The receiver Bob:
  - The channel provides Bob with one of the messages $y_1, \ldots, y_M$.
  - The probability of receiving message y_n is denoted by q_n.

To quantify the amount of information transmitted between sender and receiver in this scenario we describe Alice's messages by a random variable X which can take the values $x_1, \ldots, x_N$. X takes on the value x_j with probability[1] p_j and the probabilities sum to one:

$$\sum_j p_j = 1.$$

The amount of information contained in a message, or equivalently in the random variable X, is defined as the number of bits which are on average at least required to store an outcome of a perfect measurement of X. It tells us how much we learn from reading message X.

When measuring X, the uncertainty about its content is reduced. The information gained about the message is therefore defined as the reduction in information content induced by a measurement. After a perfect measurement of X we know for sure which message was sent, and subsequent measurements on this message will not tell us anything new. This process thus reduces the information content to zero, and the gained information equals the original information content. For the general case of imperfect measurements, the original information content and the gained information do not agree and some residual uncertainty about the message is left. In the following, we discuss these properties in relation to the sender, receiver, and communication channel, respectively.

13.1.2 Shannon's noiseless coding theorem

The information content of a message sent out by Alice is given by the *Shannon entropy* $H(X)$, defined as

$$H(X) = -\sum_{j=1}^{N} p_j \log_2(p_j).$$

This definition only makes sense if the number of messages sent out by Alice tends to infinity. The Shannon entropy $H(X)$ does not depend on the values x_j of X but just on the probabilities p_j. It is thus applicable to any kind of message if the probability distribution of the messages is known. Shannon's noiseless coding theorem states that $H(X)$ is indeed the smallest possible average number of bits required to store one of Alice's messages. In other words, a message x_j can on average be compressed to $H(X)$ bits using an optimal code for message encoding. We are not going to prove the coding theorem (see e.g. Nielsen and Chuang (2000), chapter 12 for a proof), but rather illustrate it using the following qualitative discussion and examples.

A message which occurs with probability zero does not add to $H(X)$, since

$$\lim_{p \to 0} p \log_2(p) = 0.$$

This is sensible as we cannot gain any information from messages which are never sent and no bits need to be stored for unsent messages. If only one message appears with certainty it also does not contain any information since $1 \log_2 1 = 0$. As the number of

---

[1] We assume that the sequence of messages is independent and identically distributed (i.i.d.), that is, p_j does not depend on other messages in the sequence.

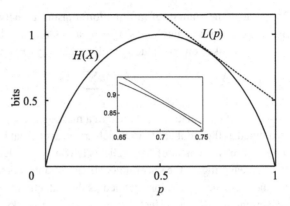

Fig. 13.1 The solid curve shows the Shannon entropy $H(X)$ as a function of probability p for two messages. Message A is sent with probability p and message B with probability $1 - p$. A maximum of $H(X)_{max} = 1$ bit is reached at $p_{max} = 1/2$. The dashed curve shows the average bit string length $L(p)$ for the compression scheme described in the text. The inset zooms into the region where $L(p)$ and $H(X)$ become similar and shows that the encoding is never optimal.

ever-repeated messages tends to infinity, the number of bits required to store it will go to zero, no matter what the content and type of message. Nothing new can be learnt from repeatedly measuring one message only since the outcome can be predicted with certainty. The Shannon entropy is bounded by $0 \leq H(X) \leq \log_2(N)$ with $H(X) = \log_2(N)$ if and only if (iff) $p_j = 1/N \; \forall \; j$, that is all possible messages appear with the same probability.[2]

We assume that Alice sends two messages A and B. She chooses A with probability $p_A = p$ and B with probability $p_B = 1 - p$. Using the above expression for the Shannon entropy we find

$$H(X) = -p \log_2(p) - (1 - p) \log_2(1 - p).$$

The maximum information content is reached when

$$\frac{dH(X)}{dp} = -\log_2(p) + \log_2(1 - p) = 0,$$

which yields $p_{max} = 1/2$ and minima are obtained for $p_{min,1} = 0$ and for $p_{min,2} = 1$ since the possible range of values of p is bounded by $0 \leq p \leq 1$. At p_{max} each message contains one bit of information, while at p_{min} the information content is zero as expected. The shape of $H(X)$ is shown in Figure 13.1 as a function of p.

The simplest encoding is to identify message A with bit value 0 and message B with bit value 1. Each of them then requires one bit independently of the value of p. This encoding is optimal for $p = 1/2$ but not otherwise. We now wish to improve on this simple encoding of Alice's messages, considering the case $p > 1/2$. We choose to encode AA as 0, AB as 10, and B as 11. No extra "start/stop" bits indicating the beginning or ending of a message are necessary since it is understood that bit strings starting with 1 have a length of two bits while a 0 indicates a bit string of length one bit.[3] The message string AA occurs with

---

[2] The symbol $\forall$ means "for all."

[3] Since we consider the case where the number of messages tends to infinity, there is no need to address the question of how to initiate the communication.

probability $p_{AA} = p^2$, AB with $p_{AB} = p(1 - p)$, and B with $p_B = 1 - p$ (see Exercise 13.1). If the number of substrings to be encoded is N_s, then the number of bits in the encoded string will on average be $N_e = p_{AA}N_s + 2p_{AB}N_s + 2p_BN_s$ and these N_s encodings will on average have encoded $N_o = 2p_{AA}N_s + 2p_{AB}N_s + p_BN_s$ messages. The average length of an encoded message is therefore given by

$$L(p) = \frac{N_e}{N_o} = \frac{2 - p^2}{1 + p}.$$

The comparison of the average length $L(p)$ with the Shannon entropy $H(X)$ shown in Figure 13.1 reveals that this encoding is never optimal. It is only better than the simple encoding, $L(p) < 1$, when $p > (\sqrt{5} - 1)/2$. The bit values 0 and 1 appear with the same probability in the encoded string for $p = \sqrt{3} - 1$, and hence this example shows that while optimally encoded bit strings contain 0 and 1 with the same probability, the converse is not true.[4]

This example illustrates that finding an optimal encoding for given message probabilities is a non-trivial task, as is also indicated by the varying performance of data compression software. However, optimal compression procedures can easily be found for some specific probability distributions. We consider the case where Alice sends three different messages X with possible values A, B, and C. The probability for A is $p_A = 1/2$, while B, C have probabilities $p_B = p_C = 1/4$. The Shannon entropy is given by

$$H(X) = -p_A \log_2 p_A - p_B \log_2 p_B - p_C \log_2 p_C = \frac{\log_2 2 + \log_2 4}{2} = \frac{3}{2}.$$

By encoding A using the bit string 0, B as 10 and C as 11 the average length L of a bit string representing a value of X is $L = 3/2 = H(X)$. This code is optimal according to the coding theorem; the messages cannot be compressed further without losing information. Note that the bit string of optimally encoded messages contains 0 and 1 at each position with the same probability of $1/2$.

The noiseless coding theorem quantifies the amount of information contained in the messages sent out by Alice. It is thus also often called *Shannon's source coding theorem*. When a message Y is received by Bob it will most often have been subject to noise and thus not be identical to X. Therefore the question arises what Bob learns about X when reading the received messages Y.

13.2 Mutual information

The message Y received by Bob via the communication channel takes on values y_n with probabilities q_n. Bob reads Y and gains information $H(Y)$. We wish to quantify the amount of information this reveals about the original message X. Thus we first introduce the joint

---

[4] This can be explained by noting that encoding may introduce correlations between different bits of the encoded string, for example, consider the encoding $A \to 01, B \to 10$.

entropy of the variables X and Y as

$$H(X, Y) = - \sum_{j,n} p(x_j, y_n) \log_2 (p(x_j, y_n)),$$

where $p(x_j, y_n)$ is the probability that X takes on value x_j *and* Y takes on value y_n. The marginal probability distributions are given by $p_j = \sum_n p(x_j, y_n)$ and $q_n = \sum_j p(x_j, y_n)$. The joint entropy is the total information content of variables X and Y.

Furthermore, we define the entropy of X conditional on knowing Y by

$$H(X|Y) = H(X, Y) - H(Y).$$

$H(X|Y)$ tells us how uncertain we are about the value of X after reading Y. If X and Y are uncorrelated, that is the probability q_n of Y taking on the value y_n is independent of the value of X, then we expect Bob to learn nothing about X when measuring Y. In this case

$$p(x_j, y_n) = p_j q_n,$$

and we find

$$H(X, Y) = H(X) + H(Y).$$

This gives $H(X|Y) = H(X)$ and therefore Bob's uncertainty about message X has not changed through reading message Y, consistent with what we expected. However, if the value of X is fixed by measuring Y, that is[5]

$$p(x_j, y_n) = \begin{cases} q_n & \text{for} \quad j = n \\ 0 & \text{otherwise} \end{cases},$$

we find that $H(X, Y) = H(Y)$ and $H(X|Y) = 0$. When Bob measures Y he gains information $H(Y)$ and thus reduces the uncertainty of the combined messages X and Y to zero, allowing him to infer the value of X. In other words, if Bob measures message y_n he is certain that message x_n was sent by Alice.

In a system that obeys local realism, messages exist irrespective of whether they are measured or not, and reading Y does not affect X. In this case we can always write $p(x_j, y_n) = p(x_j|y_n)p(y_n)$, where $p(x_j|y_n)$ is the conditional probability of Alice having sent message x_j given that Bob has read message y_n. This is a probability distribution with $p(x_j|y_n) \geq 0$ and $\sum_j p(x_j|y_n) = 1$, $\forall n$. We thus find that $H(X|Y) = - \sum_{j,n} p(x_j, y_n) \log_2 p(x_j|y_n) \geq 0$. The conditional entropy will only be zero if X and Y are perfectly correlated, meaning that $p(x_j|y_n)$ only takes on values 0 or 1. Furthermore, it can be shown that $H(X|Y) \leq H(X)$, which means that reading Y never increases uncertainty about the message X in a local realistic system.

Based on these observations we introduce the *mutual information* of X and Y as

$$H(X : Y) = H(X) + H(Y) - H(X, Y).$$

This can also be written as $H(X : Y) = H(X) - H(X|Y) = H(Y) - H(Y|X)$, which shows that the mutual information fulfills $H(X : Y) \geq 0$, $H(X : Y) \leq H(X)$ and $H(X : Y) \leq$

---

[5] Here we assume, for simplicity, that $M = N$ and that the ordering of messages is preserved in the transmission. No messages are lost or created between sender and receiver.

$H(Y)$. We find that no mutual information is contained in X and Y if they are uncorrelated, since then $H(X, Y) = H(X) + H(Y)$. In this case no information is transmitted from Alice to Bob. If X and Y are perfectly correlated we have $H(X, Y) = H(Y) = H(X)$, and the mutual information between X and Y takes its maximal value $H(X : Y) = H(X)$. The information sent by Alice can be completely restored at the receiver side in this case, consistent with $H(X|Y) = 0$.

13.3 The communication channel

We can now quantify the communication channel in terms of its channel capacity. This defines the amount of information which is transmitted by the channel in a single use. Note that it is not important that the messages arrive without being altered. The only criterion is whether Bob can reconstruct Alice's message from the output. For instance, for a channel which maps $1 \to 1$ and $0 \to 0$ we find $H(X : Y) = H(X)$ and a channel with $1 \to 0$ and $0 \to 1$ also has $H(X : Y) = H(X)$. However, a channel with $1 \to 1$ and $0 \to 1$ does not transmit information, as can be seen by working out $H(X : Y) = 0$.

A *noiseless channel* $\mathcal{N}$ produces messages Y which are perfectly correlated with the initial messages X, and thus $H(X : Y) = H(X)$. In this case Alice can transmit $H(X)$ bits of information with every use of $\mathcal{N}$. By exploiting Shannon's noiseless coding theorem, $H(X) = \log_2(N)$ can be achieved. The channel capacity $C(\mathcal{N})$ is therefore given by

$$C(\mathcal{N}) = \log_2(N).$$

If a *noisy channel* is used, X and Y will be correlated but not perfectly. The question then is whether by redundant encoding one can ensure arbitrarily good reliability of the channel for the reduced amount of information sent via such an encoding. *Shannon's noisy channel coding theorem* states that this is possible with a channel capacity of

$$C(\mathcal{N}) = \max_{\{p_j\}}\{H(X : Y)\},$$

where the maximum is taken over all possible input probability distributions p_j of X.

Example 13.1 A noisy channel can transmit two messages A and B. Message A is always transmitted faithfully but message B is only transmitted correctly with probability $1 - \ell$ while it is turned into message A otherwise.

We work out the mutual information established by each use of the channel for $p_A = p$ and $p_B = 1 - p$ as a function of p. The joint probabilities are given by $p(A, A) = p$, $p(A, B) = 0$, $p(B, A) = (1 - p)\ell$, $p(B, B) = (1 - p)(1 - \ell)$, from which the joint entropy $H(X, Y) = -\sum_{x,y \in \{A,B\}} p(x, y) \log_2(p(x, y))$ follows. The sender's Shannon entropy is $H(X) = -\sum_{x \in \{A,B\}} p_x \log_2(p_x)$ while the receiver's entropy is given by $H(Y) = -\sum_{y \in \{A,B\}} q_y \log_2(q_y)$, where $q_A = p + (1 - p)(1 - \ell)$ and $q_B = (1 - p)\ell$. For a general ℓ the mutual information takes on a maximum for $p_{\max} = (\ell^{\ell/(\ell-1)} - \ell)/(1 - \ell + \ell^{\ell/(\ell-1)})$.

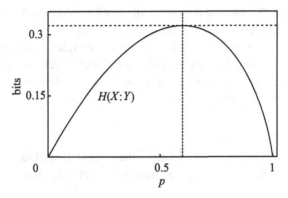

Fig. 13.2   Mutual information created by a lossy channel with $\ell = 1/2$ as a function of encoding p. The dashed grid lines indicate the value $p_{max} = 3/5$ and the achieved maximum of $H(X : Y)$, which is the channel capacity.

Here we specialize to the case $\ell = 1/2$, for which the mutual information simplifies to

$$H(X : Y) = p + \frac{p-1}{2} \log_2 (1-p) - \frac{p+1}{2} \log_2 (p+1)$$

and is shown in Figure 13.2 as a function of p. It has a maximum for $p_{max} = 3/5$, at which it takes a value of $H(X : Y)_{max} = C(\mathcal{N}) = \log_2(5) - 2 \approx 0.322$. Despite the large rate of errors ℓ this channel can still be used to faithfully transmit about a third of a bit in each use! However, this example tells us nothing about how to encode messages to achieve the channel capacity.

Remark   Sometimes channel capacity is given in bits/sec, which is $C(\mathcal{N})$ times the number of possible channel uses per second.

13.4 Connection to statistical physics

The Shannon entropy is a generalization of the equation $S = k_B \ln W$ from statistical physics, where k_B is the Boltzmann constant and W the number of microstates accessible to the system. This equation follows from our definition of the Shannon entropy by identifying each microstate with one of the messages and assuming that each of them is occupied with the same probability $1/W$. The Shannon entropy for this type of "sender" takes on its maximum value and equals S up to an unimportant constant factor. The thermodynamic entropy S is thus a measure of our ignorance of the thermal state, where all accessible microstates are equally likely to be populated.

Further reading

The Shannon entropy and its basic properties are covered in chapter 11 of Nielsen and Chuang (2000), including all relevant proofs. More advanced introductions to classical information theory can, for example, be found in chapter 5 of John Preskill's lecture notes, available online in Preskill (1997–2011), or in Vedral (2006). The basics of information theory required for classical cryptography are contained in Schneier (1995).

Exercises

13.1 Consider the encoding $AA \rightarrow 0, AB \rightarrow 10$, and $B \rightarrow 11$ as in Section 13.1.2. Starting from a string of N_o messages show that the average number of message strings used for the encoding is given by $N_{AA} = N_o p^2/(1 + p)$, $N_{AB} = N_o(1 - p)p/(1 + p))$, and $N_B = N_o(1 - p)/(1 + p)$. With the number of encoding steps given by $N_s = N_{AA} + N_{AB} + N_B$ show that the probabilities $p_{AA} = p^2$, $p_{AB} = p(1 - p)$ and $p_B = 1 - p$ given in Section 13.1.2 follow.

13.2 Alice prepares messages A, B, and C with probabilities p_A, p_B, and p_C, respectively. Show that her messages contain maximum information for $p_A = p_B = p_C = 1/3$ and work out this maximum amount of information.

13.3 Alice creates messages A, B, C, and D. She chooses them with probabilities $p_A = 1/2$, $p_B = 1/4$, and $p_C = p_D = 1/8$. How much information is contained in one of her messages? Find an optimal bit-code for encoding these messages and show that in this code each bit has equal probability of having values 0 or 1.

13.4 Now imagine that Alice uses trits (with values 0, 1, 2) instead of bits to encode her messages, and chooses to send A, B, C, D, and E with probabilities $p_A = p_B = 1/3$ and $p_C = p_D = p_E = 1/9$. What is an optimal code in this case? Show that in the optimal encoding each trit has equal probability of having values 0, 1, and 2.

13.5 Show that the conditional entropy $H(X|Y)$ is always larger than or equal to zero if local realism is assumed.

13.6 A communication channel transmits two messages A and B. With probability ℓ the two messages are swapped on the channel but otherwise transmitted faithfully. Calculate the channel capacity, making use of its symmetry, and show that the channel is ideal for $\ell = 0$ and $\ell = 1$.

Quantum information

We first revise the description of a quantum mechanical system in terms of density operators based on our initial discussion of these topics in Part I and introduce the notion of global and local measurements. This is then followed by a discussion of quantum information concepts similar to those introduced for classical systems.

14.1 The density operator

All quantum mechanical expectation values for a system in the pure state $|\psi\rangle$ can be rewritten in terms of a trace

$$\langle M \rangle = \langle \psi | M | \psi \rangle = \mathrm{Tr}\, \{|\psi\rangle \langle \psi | M\},$$

with M the observable of interest. We can thus define the operator $\rho = |\psi\rangle \langle \psi|$ and use it to replace the state vector $|\psi\rangle$. This operator is called the density operator, or simply the state of the quantum system. All observable quantities can be worked out from the density operator.[1]

Sometimes the state vector of a system is not known, but it is known that the system will be described by a state vector $|\psi_j\rangle$ with probability p_j. This can happen for a number of reasons, for example (i) if the system preparation is imperfect, like for a thermal state; (ii) after a measurement if the outcome is not revealed to the observer; (iii) through decoherence processes; or (iv) when it produces quantum messages $|\psi_j\rangle$ with probability p_j. In these cases the expectation value of an operator has to be worked out for each possible state vector and to be weighted with the classical probability[2] p_j. This yields

$$\langle M \rangle = \sum_j p_j \mathrm{Tr}\, \{|\psi_j\rangle \langle \psi_j | M\}.$$

The trace is linear and therefore defining the general density operator

$$\rho = \sum_n p_j |\psi_j\rangle \langle \psi_j|$$

allows us to write $\langle M \rangle = \mathrm{Tr}\, \{\rho M\}$. Using a density operator to describe a quantum system thus permits a compact and efficient way to include classical uncertainty about its state

---

[1] The unmeasurable global phase of a state vector is gone.
[2] This is a classical uncertainty about which state vector describes the quantum system and is fundamentally different from quantum uncertainty arising from non-commuting observables.

vector into quantum mechanical calculations. The state of a quantum system is said to be *pure* if only one state vector contributes to it, that is $\rho = |\psi\rangle\langle\psi|$ and it is *mixed* otherwise. Iff a state is pure $\mathrm{Tr}\left\{\rho^2\right\} = 1$.

14.1.1 The density matrix

For a given basis $|\phi_n\rangle$ the density operator ρ is fully specified by the matrix elements $\rho_{nm} = \langle\phi_n|\rho|\phi_m\rangle$, as can be seen by using the completeness of the basis $\sum_n |\phi_n\rangle\langle\phi_n| = \mathbb{1}$. Introducing this closure relation to the left and right of the density operator, we find

$$\rho = \mathbb{1}\rho\mathbb{1} = \sum_{n,m} |\phi_n\rangle\langle\phi_n|\,\rho\,|\phi_m\rangle\langle\phi_m| = \rho_{nm}\,|\phi_n\rangle\langle\phi_m|\,.$$

The terms density operator and density matrix are thus often used interchangeably. While it is *necessary* to specify the basis in which the density matrix is written, this is often understood to be the computational basis and not stated explicitly.

14.1.2 Entangled states

A pure quantum state $|\psi\rangle$ defined in Hilbert space $\mathcal{H} = \mathcal{H}_A \otimes \mathcal{H}_B$ is entangled iff it is not a product state, that is if it cannot be written as $|\psi\rangle = |\psi\rangle_A \otimes |\psi\rangle_B$, and this criterion is usually easily checked. For a mixed state ρ to be entangled it must *not* be possible to write it in the following form:

$$\rho = \sum_j p_j \rho_A^{(j)} \otimes \rho_B^{(j)},$$

with p_j the probability of finding the system in state $\rho_A^{(j)} \otimes \rho_B^{(j)}$. This is in many cases not straightforward to check, since there are usually many possible ways to decompose a mixed state.[3] For instance, when considering the density matrix in the computational basis

$$\rho_1 = \frac{1}{4}\begin{pmatrix} 1 & 0 & 0 & -1 \\ 0 & 1 & -1 & 0 \\ 0 & -1 & 1 & 0 \\ -1 & 0 & 0 & 1 \end{pmatrix},$$

it may not be immediately obvious that this matrix corresponds to the density operator $\rho_1 = (|+\rangle\langle+| \otimes |-\rangle\langle-| + |-\rangle\langle-| \otimes |+\rangle\langle+|)/2$ and is therefore not entangled. Mixed states which violate the conditions imposed on conditional entropy and mutual information by local realism are entangled, while the opposite is not necessarily true.

---

[3] For low-dimensional systems the partial transpose method provides an easy way to check for entanglement of mixed states.

14.2   Global and local measurements

In undergraduate quantum mechanics the measurement of a Hermitian operator M with eigenvectors $|\phi_j\rangle$ and non-degenerate eigenvalues a_j is introduced. In a single measurement the eigenvalue a_j is obtained and the system collapsed into the state $|\phi_j\rangle$ with probability $p_j = |\langle\phi_j|\psi\rangle|^2 = \langle\phi_j|\psi\rangle\langle\psi|\phi_j\rangle$. The average measurement outcome for a system in a pure state $|\psi\rangle$ is therefore given by $\langle M\rangle = \langle\psi|M|\psi\rangle$. For a mixed state ρ this extends using identical arguments as above. Outcome a_j is obtained with probability $p_j = \langle\phi_j|\rho|\phi_j\rangle$. If outcome a_j is obtained, the mixed state is collapsed into $|\phi_j\rangle\langle\phi_j|\rho|\phi_j\rangle\langle\phi_j|/p_j = |\phi_j\rangle\langle\phi_j|$.

For an operator with degenerate eigenvalues a_j and eigenvectors $|\phi_{j,n}\rangle$, where n enumerates the set of degenerate eigenvectors for eigenvalue a_j, the probability of outcome a_j is given by $p_j = \sum_n \langle\phi_{j,n}|\rho|\phi_{j,n}\rangle$. For outcome a_j the mixed state is collapsed into

$$\rho^{(j)} = \frac{1}{p_j}\sum_{n,m}|\phi_{j,n}\rangle\langle\phi_{j,n}|\rho|\phi_{j,m}\rangle\langle\phi_{j,m}|.$$

Note that p_j is included here to obtain a normalized state. The observer cannot distinguish states with degenerate eigenvalues from his measurements. The degrees of freedom that could distinguish these states are not accessible.

In many cases we will not be interested in the eigenvalues a_j obtained in a measurement of an operator M. We instead specify a measurement via a set of orthonormal states $|\phi_j\rangle$ and ask about the probability p_j of projecting the system into state $|\phi_j\rangle$. The corresponding eigenvalues are then assumed to be non-degenerate, so that these states are distinguishable for the observer. The operator M and its eigenvalues a_j are not required to calculate the probabilities p_j. Nevertheless, a corresponding observable M could be constructed as $M = \sum_j a_j|\phi_j\rangle\langle\phi_j|$ with arbitrarily chosen real $a_j \neq a_l$ for $j \neq l$.

In quantum information we are usually concerned with the situation where the system consists of sender Alice and receiver Bob, spatially separated. The situation is described by a Hilbert space $\mathcal{H} = \mathcal{H}_A \otimes \mathcal{H}_B$, where $\mathcal{H}_A$ describes the degrees of freedom accessible to Alice and $\mathcal{H}_B$ those that Bob can measure. In order to proceed it is useful to introduce the partial trace of an operator, which is the trace over a subspace of the total Hilbert space.

For instance, by tracing over the Hilbert space pertaining to Alice's degrees of freedom we will be left with an operator that only acts on the degrees of freedom accessible to the receiver Bob. We write the basis of $\mathcal{H}$ as $|\phi_{i,j}\rangle = |\phi_i\rangle_A \otimes |\phi_j\rangle_B$, where $|\phi_i\rangle_A$ describes Alice's degrees of freedom and $|\phi_j\rangle_B$ those of the receiver. The partial trace of an operator M over subspace $\mathcal{H}_A$ is then given by

$$\mathrm{Tr}_A\{M\} = \sum_i {}_A\langle\phi_i|M|\phi_i\rangle_A = \sum_{i,j,l}|\phi_j\rangle_B M_{ij,il}\,{}_B\langle\phi_l|.$$

Here we have used the notation $M_{ij,kl} = \langle\phi_{i,j}|M|\phi_{k,l}\rangle$ and the rule[4] ${}_A\langle\phi_i|\phi_{k,l}\rangle = \delta_{ik}|\phi_l\rangle_B$. The resulting operator contains only basis elements corresponding to degrees of freedom

---

[4] This is not a proper scalar product and slightly sloppy, though widely used, notation.

of Bob; it only acts on Bob's subspace. We have thus *traced out* the sender Alice. Note that subsequently tracing over Bob yields the trace of the operator.

Any observable local to Alice can be written as $M = M_A \otimes \mathbb{1}_B$ since Alice cannot directly measure degrees of freedom located at the receiver site. We work out the expectation value of M by first tracing over Bob and then over Alice, and find

$$\langle M \rangle = \mathrm{Tr}_A \{\mathrm{Tr}_B \{M\rho\}\} = \mathrm{Tr}_A \{M_A \mathrm{Tr}_B \{\mathbb{1}_B \rho\}\} = \mathrm{Tr}_A \{M_A \rho_A\},$$

where $\rho_A = \mathrm{Tr}_B \{\rho\}$ is the reduced density operator of the sender. All information about the quantum system which can be measured locally by Alice is contained in this reduced density operator ρ_A. The same argument holds for the receiver, who only has access to information contained in the reduced density operator $\rho_B = \mathrm{Tr}_A \{\rho\}$.

The measurement of a local observable $M = M_A \otimes \mathbb{1}_B$ can be described like a global measurement above by specifying the eigenstates of M_A, written as $|\phi_j\rangle_A$, and writing the global observable as $M = \sum_{j,n} a_j |\phi_{j,n}\rangle\langle\phi_{j,n}|$ with a_j the (assumed) non-degenerate eigenvalues of M_A and $|\phi_n\rangle_B$ an arbitrarily chosen basis of Bob's subspace[5] $\mathcal{H}_B$. The operator M has degenerate eigenvalues,[6] since Alice cannot measure Bob's degrees of freedom. The measurement collapses the state into $|\phi_j\rangle_A$ and the total system into $|\phi_j\rangle_A\langle\phi_j| \otimes \rho_B^{(j)}$ with probability $p_j = {}_A\langle\phi_j|\rho_A|\phi_j\rangle_A$ and

$$\rho_B^{(j)} = \sum_{n,m} q_{j,n,m} |\phi_n\rangle_B\langle\phi_m| \qquad \text{where} \qquad q_{j,n,m} = \frac{1}{p_j} \langle\phi_{j,n}|\rho|\phi_{j,m}\rangle.$$

If Alice informs Bob about her measurement outcome then Bob knows that his reduced density operator is $\rho_B^{(j)}$. Such communication, which can be classical, thus reduces Bob's uncertainty about his state, changing it from ρ_B to $\rho_B^{(j)}$.

Example 14.1 We consider the case where Alice and Bob share a Bell state $|\psi^-\rangle$. In this case the Hilbert space is split into $\mathcal{H} = \mathbb{C}^2 \otimes \mathbb{C}^2$ and the computational basis states are $|jn\rangle = |j\rangle_A \otimes |n\rangle_B$ with $j, n = 0, 1$. The reduced density operator of Bob is

$$\rho_B = {}_A\langle 0| \psi^-\rangle\langle\psi^- | 0\rangle_A + {}_A\langle 1| \psi^-\rangle\langle\psi^- | 1\rangle_A = \frac{1}{2} \left(|0\rangle_B\langle 0| + |1\rangle_B\langle 1|\right) = \frac{\mathbb{1}_B}{2}.$$

By symmetry we obtain $\rho_A = \mathbb{1}_A/2$. Despite the overall state being pure, the Bell state reduces to a maximally mixed state for Alice and also for Bob.

The probabilities for a local measurement at Alice's site to project the system into the orthogonal states $|\phi_1\rangle = \alpha|0\rangle + \beta|1\rangle$ and $|\phi_2\rangle = \alpha^*|1\rangle - \beta^*|0\rangle$ are $1/2$ irrespective of α and β. Depending on the measurement outcome, the system is either projected into the pure state $|\phi_1\rangle_A|\phi_2\rangle_B$ or the pure state $|\phi_2\rangle_A|\phi_1\rangle_B$, giving perfectly orthogonal states irrespective of the local measurement and its outcome. When averaging over many measurements, or if the outcome is unknown (e.g. in a decoherence process), the resulting mixed state is

$$\rho = \frac{1}{2}(|\phi_1\rangle_A\langle\phi_1| \otimes |\phi_2\rangle_B\langle\phi_2| + |\phi_2\rangle_A\langle\phi_2| \otimes |\phi_1\rangle_B\langle\phi_1|).$$

---

[5] Note that $\sum_n |\phi_n\rangle_B\langle\phi_n|$ is the identity operator on the receiver subspace independently of the choice of basis.
[6] Except for the trivial case where the receiver only consists of a one-dimensional Hilbert space.

If instead we consider the system to initially be in the non-entangled mixed state

$$\rho = \frac{1}{2} \left(|01\rangle \langle 01| + |10\rangle \langle 10| \right),$$

then we again get maximally mixed reduced density operators but now local measurements by Alice will only lead to perfectly orthogonal states at each site if she chooses to measure in the computational basis. For a measurement outcome $|\phi_1\rangle_A$ the resulting state will in general be mixed and given by

$$\rho = \frac{1}{2} (|\phi_1\rangle_A \langle \phi_1| \otimes \left(|\beta|^2 |0\rangle_B \langle 0| + |\alpha|^2 |1\rangle_B \langle 1| \right).$$

For instance, if Alice measures in the X-basis then Bob's state after the measurement will be maximally mixed, irrespective of Alice's measurement outcome, and whether she communicates it to Bob or not.

14.3 Information content of a density operator

We are now in a position to extend the notion of measures of information to the quantum case. Classical messages X are replaced by the density operator. Alice prepares quantum states $|x_j\rangle$ to be sent to Bob with probability p_j. These encode the messages and are compactly described by a density operator

$$\rho = \sum_j p_j |x_j\rangle \langle x_j| ,$$

which is Hermitian. For simplicity we assume here that the messages $|x_j\rangle$ are orthogonal and normalized eigenvectors of ρ.[7]

The von Neumann entropy of the state ρ is defined by

$$S(\rho) = -\text{Tr} \left\{ \rho \log_2 (\rho) \right\} = - \sum_j p_j \log_2 (p_j) .$$

It is a measure of our ignorance about the quantum state and plays a similar role for quantum states as the Shannon entropy does for classical random variables. Like in the classical case it only depends on the probabilities with which a message is sent, but not on the actual quantum messages $|x_j\rangle$. Using the von Neumann entropy, quantum states can almost be treated as if they were information. Furthermore, up to a constant factor, it reduces to the entropy introduced in quantum statistical mechanics for density operators.

14.3.1 Schumacher's quantum noiseless channel coding theorem

In analogy to Shannon's noiseless coding theorem, Schumacher showed that states ρ in a d-dimensional Hilbert space $\mathcal{H}_A$ can be compressed. In particular, it is possible to reliably

---

[7] Since ρ is Hermitian such a representation can always be found even if Alice chooses to prepare non-orthogonal messages.

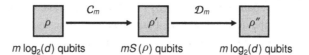

Fig. 14.1 Compression $\mathcal{C}_m$ and decompression $\mathcal{D}_m$ of m copies of a quantum state ρ. The size of the original Hilbert space corresponds to $m \log_2(d)$ qubits, which are compressed to $mS(\rho)$ qubits.

compress and decompress ρ to a quantum state in a Hilbert space $\mathcal{H}_R$ with dimension

$$\dim(\mathcal{H}_R) = 2^{S(\rho)},$$

and can thus be viewed as being represented by $S(\rho)$ qubits. Like in classical encoding this only works on average, that is if the source produces a large number m of quantum messages in the limit $m \to \infty$. The procedure is shown schematically in Figure 14.1. Reliably in this case means that the entanglement of the original state with its environment is preserved after compression $\mathcal{C}_m$ and decompression $\mathcal{D}_m$. This is quantified by the entanglement fidelity, which for reliable encoding tends to 1 for large m.

14.4 Joint entropy and mutual information

We define the joint entropy for the quantum state of Alice and Bob, ρ_{AB}, in analogy to classical information theory as

$$S(\rho_{AB}) = -\mathrm{Tr}\left\{\rho_{AB} \log_2(\rho_{AB})\right\}.$$

The reduced density operators ρ_A for Alice and ρ_B for Bob yield the corresponding information contents $S(\rho_A)$ and $S(\rho_B)$. The conditional entropy and mutual information are defined in analogy to the classical case:

$$S(\rho_A|\rho_B) = S(\rho_{AB}) - S(\rho_B),$$

$$S(\rho_A : \rho_B) = S(\rho_A) + S(\rho_B) - S(\rho_{AB}).$$

These quantities replace those introduced in classical information theory but do not have the classically expected properties. For instance, the conditional entropy can become negative as we will explicitly work out later for the case of two entangled qubits.

In contrast to the classical case Bob cannot read his messages without affecting the quantum state. We investigate measurements on one qubit initially prepared in the pure state $|+\rangle$, which is measured in the computational basis. We study two different scenarios: (i) the outcome of the measurement is $|0\rangle$ and (ii) the outcome of the measurement is not revealed. Case (ii) can happen in a decoherence process where the measurement is performed by the environment and the outcome is not accessible to an experimentalist. How does the information content of the qubit change in the two cases?

(i) The initial state is $\rho_i = |+\rangle\langle+|$. This is a pure state (state $|+\rangle$ is prepared with certainty) and thus has entropy $S(\rho_i) = 0$. After the measurement the state is $\rho_f = |0\rangle\langle0|$, which is again pure and so $S(\rho_f) = 0$. While the quantum state changes in the measurement process the qubit remains in a pure quantum state and never contains any information. We note that the two possible measurement outcomes $|0\rangle$ and $|1\rangle$ occur with probability $1/2$ each, despite measuring a pure state containing no information. Instead, this probabilistic measurement outcome originates from the quantum mechanical incompatibility of the measurement basis and the quantum state.

(ii) The actual outcome of the measurement is unknown and it therefore produces quantum messages $|0\rangle$ or $|1\rangle$ with probability $1/2$ each. The corresponding mixed state is

$$\tilde{\rho}_f = \frac{1}{2} |0\rangle \langle 0| + \frac{1}{2} |1\rangle \langle 1| .$$

This state has entropy $S(\tilde{\rho}_f) = 1$. Lacking knowledge of the measurement outcome turns the initial pure state into a mixed state. Its information content – the uncertainty about the qubit state – increases.

Finally we consider what happens if a measurement in the computational basis is performed on the state $\tilde{\rho}_f$ and the outcome is $|0\rangle$. In this process the entropy is reduced from $S(\tilde{\rho}_f) = 1$ to $S(|0\rangle \langle 0|) = 0$. The mixed state is turned into a pure state by the measurement, and one bit of information about the original state $\tilde{\rho}_f$ is gained. Only this type of measurement is in line with what is understood by reading a message in the classical setting.

14.5 Quantum channels

A quantum channel transforms an input state ρ into an output state[8] $\rho_{\text{out}} = T(\rho)$. Mathematically the channel is described by a completely positive, linear unital map T which acts according to

$$T(\rho) = \sum_{j=1}^{n} E_j \rho E_j^{\dagger}.$$

Here E_j are the so-called *Kraus operators* which fulfill $\sum_j E_j^{\dagger} E_j \leq \mathbb{1}$. The case $T = \mathbb{1}$ describes an ideal channel. The channel capacity quantifies the number of qubits which can faithfully be transmitted, and is fully understood only for special cases. For instance the Holevo–Schumacher–Westmoreland (HSW) theorem gives the channel capacity if only product input states are used. A more detailed discussion of channel capacity is beyond the scope of this text. Instead, we consider two simple examples of how noise and decoherence affect the information content of quantum systems.

14.5.1 Dephasing channel

In this channel a qubit is transmitted faithfully with probability $1 - \ell$ and undergoes a phase flip error σ_z with probability ℓ. A phase flip error can, for example, be caused by

---

[8] ρ and $T(\rho)$ can in principle live in different Hilbert spaces $\mathcal{H}_1$ and $\mathcal{H}_2$, but here we will assume $\mathcal{H}_1 = \mathcal{H}_2$.

fluctuating fields which affect the energies of the two qubit states differently. The channel is described by the Kraus operators $E_1 = \sqrt{1 - \ell}\mathbb{1}$ and $E_2 = \sqrt{\ell}\sigma_z$. We consider the action of this channel on the density matrix in the computational basis and find

$$T(\rho) = \begin{pmatrix} \rho_{00} & (1 - 2\ell)\rho_{01} \\ (1 - 2\ell)\rho_{10} & \rho_{11} \end{pmatrix}.$$

The dephasing channel affects the off-diagonal elements of the density matrix, which describe the coherence between the two computational states. For $\ell = 1/2$ the coherences are completely destroyed and any phase information is lost. The diagonal matrix elements which correspond to the populations of the computational basis states are left untouched by this channel. A bit encoded in states $|+\rangle$ and $|-\rangle$ is adversely affected by this channel, while a bit stored in the computational basis states is not.

Example 14.2 We consider a photonic channel which is used to transmit two messages: $|0\rangle \equiv$ no photon present and $|1\rangle \equiv$ one photon present. The transmission of the photon causes phase fluctuations which lead to a σ_z error with probability ℓ. We investigate the following classical and quantum scenarios for using this channel to establish mutual information between sender Alice and receiver Bob.

(i) Classical messages X with values $|0\rangle$ and $|1\rangle$ are sent with probabilities $p_0 = p_1 = 1/2$ and received as messages Y with possible values $|0\rangle$ and $|1\rangle$. How much mutual information is created in one use of the channel?

Since the bit is never flipped in this channel the joint probabilities are $p(0, 0) = p(1, 1) = 1/2$, $p(0, 1) = p(1, 0) = 0$, which gives $H(X, Y) = 2$. Furthermore, $H(X) = H(Y) = 1$ and hence the mutual information is $H(X : Y) = 1$, showing that this classical encoding of a bit is not affected by phase noise.

(ii) Alice creates a two-photon entangled state $|\Psi^-\rangle$ and sends one of the photons to Bob via the channel.

Applying the dephasing channel to one of the qubits, we find

$$\rho_{AB} = T(|\Psi^-\rangle\langle\Psi^-|) = (1 - \ell)|\Psi^-\rangle\langle\Psi^-| + \ell|\Psi^+\rangle\langle\Psi^+|.$$

Since the states $|\Psi^-\rangle$ and $|\Psi^+\rangle$ are orthogonal, this density operator has eigenvalues $1 - \ell$, ℓ, 0 and 0, giving an entropy of $S(\rho_{AB}) = -\ell \log_2 \ell - (1 - \ell)\log_2(1 - \ell)$. The reduced density matrices are maximally mixed so that $S(\rho_A) = S(\rho_B) = 1$. For $\ell = 1/2$ we find $S(\rho_A : \rho_B) = 1$ and $S(\rho_A|\rho_B) = 0$. However, for all other values of ℓ the channel establishes a mutual information of $S(\rho_A : \rho_B) > 1$ and a negative conditional entropy $S(\rho_A|\rho_B) < 0$. This is not possible in any classical scenario where one bit is communicated between sender and receiver. In contrast to (i) above the quantum case is strongly affected by phase noise, which indicates that it is crucial to keep coherence in the transmission process in order to achieve mutual information larger than classically possible.

14.5.2 Amplitude damping channel

We now consider a channel where the state $|0\rangle$ is transmitted faithfully but $|1\rangle$ is transmitted correctly with probability $1 - \ell$ and decays to $|0\rangle$ with probability ℓ. The density matrix element ρ_{11} thus evolves according to $[T(\rho)]_{11} = (1 - \ell)\rho_{11}$ and the population lost in the upper state must be gained by the lower state, yielding $[T(\rho)]_{00} = \rho_{00} + \ell\rho_{11}$. A more detailed analysis shows that the coherences must evolve according to[9] $[T(\rho)]_{01} = \sqrt{1 - \ell}\rho_{01}$ to ensure that the state remains physical. The Kraus operators describing this channel are therefore given by $E_1 = |0\rangle\langle 0| + \sqrt{1 - \ell}|1\rangle\langle 1|$ and $E_2 = \sqrt{\ell}|0\rangle\langle 1|$. There is no encoding of a classical qubit that would allow it to be transmitted faithfully through this channel. For $\ell = 1$ all messages exiting the channel will be $|0\rangle$ and no information can be transmitted in this limit.

Example 14.3 We again consider the photonic channel from Example 14.2 but now assume that 20% of the photons are lost along the channel instead of the phase noise considered previously. This is an amplitude damping channel with $\ell = 1/5$. We study the two scenarios already investigated in Example 14.2.

(i) The initial state is $\rho = p|00\rangle\langle 00| + (1-p)|11\rangle\langle 11|$ and the second qubit goes through the amplitude damping channel. We immediately find $H(X) = 1$. The joint entropy can be worked out from the joint probabilities $p(0, 0) = 1/2$, $p(0, 1) = 0$, $p(1, 0) = \ell/2$, and $p(1, 1) = (1 - \ell)/2$ and gives $H(X, Y) \approx 1.361$. The probabilities on the receiver side are $q_0 = (1 + \ell)/2$ and $q_1 = (1 - \ell)/2$, and thus $H(Y) \approx 0.971$. We therefore find $H(X|Y) = H(X, Y) - H(Y) \approx 0.39$ and $H(X : Y) = H(X) - H(X|Y) \approx 0.61$. The photon loss along the channel significantly reduces the maximally achievable mutual information down from 1 bit for a lossless channel.

(ii) The quantum state after sending one qubit of the state $|\Psi^-\rangle$ down the channel is found by applying the corresponding Kraus operators. In the computational basis we find

$$\rho_{AB} = \frac{1}{2}\begin{pmatrix} \ell & 0 & 0 & 0 \\ 0 & 1 - \ell & -\sqrt{1 - \ell} & 0 \\ 0 & -\sqrt{1 - \ell} & 1 & 0 \\ 0 & 0 & 0 & 0 \end{pmatrix}.$$

This matrix is block-diagonal. The 1×1 blocks give eigenvalues $\ell/2$ and 0, and diagonalizing the remaining 2×2 block we find two further eigenvalues 0 and $1 - \ell/2$. The entropy is thus $S(\rho_{AB}) \approx 0.469$. The reduced density operator on Bob's side is $\rho_B = ((1 + \ell)|0\rangle\langle 0| + (1 - \ell)|1\rangle\langle 1|)/2$ with entropy $S(\rho_B) \approx 0.971$, while Alice has a totally mixed state $\rho_A = \mathbb{1}/2$. We thus find for the conditional entropy $S(\rho_A|\rho_B) \approx -0.502$ and a mutual information of $S(\rho_A : \rho_B) \approx 1.502$. Photon loss significantly reduces the mutual information from its maximum value of 2 for an ideal channel. The mutual information is shown in Figure 14.2 as a function of ℓ. It drops below 1, the classically allowed maximum for $\ell \geq 1/2$.

---

[9] $\rho_{10} = \rho_{01}^*$ since a density operator must always be Hermitian.

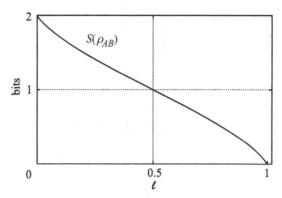

Mutual information established by an amplitude damping channel with loss probability ℓ. The dashed lines indicate the point $\ell = 1/2$ where the established mutual information drops below the limit achievable by sending one bit.

The idea of a quantum channel can even be usefully exploited to study the time evolution of physical systems. The following example demonstrates this by working out how a bit stored in internal atomic degrees of freedom degrades over time.

Example 14.4 We consider the storage of classical information in a quantum system. A bit is stored in two atomic states $|0\rangle$ and an upper state $|1\rangle$. The state $|0\rangle$ is stable while the state $|1\rangle$ is metastable and spontaneously emits photons at a rate γ. The time evolution of the atom is described by an amplitude damping channel where the probability of decay at time t is given by $\ell = 1 - e^{-\gamma t}$. How does the mutual information between the state of the atom and the initial bit change with time?

The value of the original bit is unknown and thus described by the maximally mixed state $\rho_A = (|1\rangle\langle 1| + |0\rangle\langle 0|)/2$. This contains $S(\rho_A) = 1$ bit of information. Upon writing the value of the bit into the atom, the joint bit–atom system is prepared in state $\rho_{AB} = (|11\rangle\langle 11| + |00\rangle\langle 00|)/2$. As the atom undergoes amplitude damping the density operator evolves in time according to

$$\rho_{AB}(t) = \frac{e^{-\gamma t}}{2}|11\rangle\langle 11| + \frac{\ell}{2}|10\rangle\langle 10| + \frac{1}{2}|00\rangle\langle 00| = \frac{1}{2}\begin{pmatrix} 1 & 0 & 0 & 0 \\ 0 & 0 & 0 & 0 \\ 0 & 0 & 1-e^{-\gamma t} & 0 \\ 0 & 0 & 0 & e^{-\gamma t} \end{pmatrix},$$

where the density matrix is written in the computational basis. We find the entropy of the atomic state by tracing out the initial bit, obtaining the reduced density operator

$$\rho_B = \frac{e^{-\gamma t}}{2}|1\rangle\langle 1| + \frac{2-e^{-\gamma t}}{2}|0\rangle\langle 0| = \frac{1}{2}\begin{pmatrix} 2-e^{-\gamma t} & 0 \\ 0 & e^{-\gamma t} \end{pmatrix}.$$

The entropies of these states are shown in Figure 14.3 as a function of time. Initially, any entropy solely arises from the unknown state of the bit. The uncertainty of the atomic state due to spontaneous emission then increases the overall entropy for short times $t \leq 1/\gamma$. For

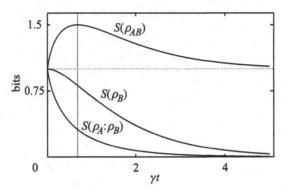

Entropies of an atom B storing a bit A. The atom undergoes spontaneous emission from its upper state at rate γ. The dashed vertical line indicates the time of maximum entropy.

times $t \gg 1/\gamma$ the atom will be in the pure state $|0\rangle$ and the entropy of the system is again solely due to the unknown state of the initial bit. However, in this process the correlation between the state of the atom and the bit is lost on a time scale $1/\gamma$, as becomes evident by looking at the mutual information $S(\rho_A : \rho_B) = S(\rho_A) + S(\rho_B) - S(\rho_{AB})$ also shown in Figure 14.3. We note that this example can equally well be described using classical information theory.

As we have seen in Examples 14.2 and 14.3, the mutual information established via distributing an entangled state can be larger than in the classical case and, as a consequence, negative conditional entropies, which are not possible in classical schemes, may arise. However, there is no (known) scheme for using solely this kind of mutual information to transmit messages from Alice to Bob. Alice initially prepares a pure state which does not contain information; each copy corresponds to the same message. Entanglement alone is thus not sufficient for communication – and faster-than-light communication is not possible. By combining entanglement and classical communication, however, improvements over conventional classical communication schemes can be achieved, for example in quantum dense coding, which we will study in the next chapter.

Further reading

The basics of quantum information theory are covered in depth in chapters 11 and 12 of Nielsen and Chuang (2000). Chapter 12 provides a comprehensive treatment of quantum channels. Again, graduate-level treatments of quantum information theory can, for example, be found in chapter 5 of John Preskill's lecture notes, available online in Preskill (1997–2011), or in Vedral (2006).

Exercises

14.1 Calculate the reduced density operator for each qubit of the Bell state $|\Psi^+\rangle$. Show that this result is the same for each Bell state and give a physical explanation. Use this to calculate the von Neumann entropy of a two-qubit system in a Bell state as well as the reduced entropies of each of the qubits separately.

14.2 The density operator of two qubits A and B is given by $\rho_{AB} = (|\Psi^-\rangle\langle\Psi^-| + |\Phi^+\rangle\langle\Phi^+| + |\Psi^+\rangle\langle\Psi^+| + |\Phi^-\rangle\langle\Phi^-|)/4$. Calculate the von Neumann entropy $S(\rho_{AB})$ and the entropies of the reduced systems $S(\rho_A)$ and $S(\rho_B)$. Is the state ρ_{AB} entangled? If it is not entangled, find the density operator ρ_{AB} in the form

$$\rho_{AB} = \sum_j p_j \rho_A^{(j)} \otimes \rho_B^{(j)}.$$

Repeat the above calculations for the state $\tilde{\rho}_{AB} = (|\Psi^-\rangle\langle\Psi^-| + |\Phi^+\rangle\langle\Phi^+|)/2$.

14.3 The density operator of a two-qubit system is given by

$$\rho_{AB} = \rho_A \otimes \rho_B.$$

Show that the von Neumann entropy of this system is given by $S(\rho_{AB}) = S(\rho_A) + S(\rho_B)$.

14.4 Show that the conditional von Neumann entropy $S(\rho_A|\rho_B)$ is equal to zero for a pure state ρ_{AB} iff ρ_{AB} is not entangled, that is, if it can be written as

$$\rho_{AB} = \rho_A \otimes \rho_B$$

and is smaller than zero for any entangled states.

14.5 Solve case (ii) of Example 14.3 but for a symmetric channel where both qubits undergo amplitude damping,[10] each with probability ℓ. This situation is realized if a central party Charlie produces the entangled state and distributes one qubit to Alice and the other to Bob.

14.6 When distributing photons the probability of a photon being lost goes exponentially with the distance L, that is $\ell = 1 - e^{-\gamma L}$. By comparing your results in Exercise 14.5 with those in Example 14.3, discuss which setup will be more suitable for the distribution of entangled pairs of photons.

[10] It may be helpful to assume that the two qubits may go through their channels in sequence when solving this problem.

15 Quantum communication

The natural choice of physical qubit for quantum communication is the photon: a photon can be transmitted quickly from the sender to a distant receiver and the technology for creating, manipulating, distributing, and measuring light pulses is well established. Many of these classical techniques can also be employed for quantum communication protocols. We will therefore focus our attention on optical setups for quantum communication.

We will study a number of optical setups in the next few sections and use the conventions introduced in Chapter 4 to analyze them. In particular, we will assume that the qubits are encoded in the degrees of freedom of a single photon. Some of the main technical challenges in quantum communication arise from this need to work with single photon pulses. For instance, photons do not interact with each other in vacuum or linear media, and interactions between two photons in non-linear optical media are relatively small, so that realizing entangling two photon gates via coherent interactions is a challenging task.

The quantum communication schemes discussed here circumvent this technical problem to a large extent. They only use non-linear optical materials for parametric down-conversion to create Bell-pairs of photons and then exploit standard linear optical devices and photo-detectors to manipulate them. No further non-linear entangling gates are required.[1]

15.1 Parametric down-conversion

In non-linear optical media a single photon can be down-converted into a pair of photons. In this coherent process the incoming photon is destroyed and two photons of lower energy are created. Energy, momentum, and angular momentum conservation rules obeyed in the process determine the correlations between the two generated photons. Since down-conversion is coherent, multiple possible paths allowed by conservation laws superpose and lead to entangled states. We consider two such situations leading to momentum (or spatial) entanglement and polarization entanglement, respectively.

15.1.1 Momentum entanglement

This can be created in non-collinear down-conversion fulfilling the phase-matching conditions as shown in Figure 15.1. Phase matching permits a photon to go down path a if and

---

[1] Single photon sources and linear optics can even be shown to be sufficient for scalable quantum computation (see further reading).

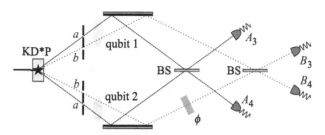

Fig. 15.1 Momentum-entangled photons generated by down-conversion in a KD*P non-linear crystal. Two pairs of paths a and b are selected by double apertures generating a Bell state of spatially encoded qubits, indicated by the ellipses. The setup after the mirror is used to measure the Bell state correlations. Coincidence clicks at the four detectors are measured after sending the photons through beam splitters (BS) as a function of the phase ϕ induced in path b_2.

only if the other photon travels along path b and vice versa. The process is coherent and thus the phase relation between the two photons is fixed, giving a Bell state of the form

$$|\Psi\rangle = \frac{1}{\sqrt{2}} \left(|a\rangle_1 |b\rangle_2 + |b\rangle_1 |a\rangle_2\right] \equiv \frac{1}{\sqrt{2}} \left(|01\rangle + |10\rangle\right),$$

where $|a\rangle_{1,2}$ and $|b\rangle_{1,2}$ denote the different photon paths and encode qubit states $|0\rangle$ and $|1\rangle$ respectively. After combining these paths using beam splitters, interference fringes are detected as a function of ϕ, indicating entanglement.

For instance, as shown in Exercise 15.1, the probability of coincidence clicks in detectors A_3 and B_3 is given by $p_{A_3,B_3} = \cos^2(\phi/2)/2$. If the down-conversion process were not coherent, then a mixed state of the form

$$\rho = \frac{1}{2} \left(|ab\rangle \langle ab| + |ba\rangle \langle ba|\right)$$

would be produced (see Exercise 15.2). The density matrix of this state lacks the off-diagonal coherences of the state $|\Psi\rangle$. One can again work out the probability of coincidence clicks in detectors A_3 and B_3 (see Exercise 15.3) and find that this is now $p_{A_3,B_3} = 1/4$, independent of the induced phase ϕ, indicating no entanglement.

The phase-matching conditions will ensure that coincidence clicks of detectors in the same path are not possible, that is $p_{A_3,A_4} = p_{B_3,B_4} = 0$. Imperfections in the setup might thus become observable through such coincidences.

15.1.2 Polarization entanglement

Non-collinear type-II down-conversion phase matching can be used to generate photons entangled in polarization. As shown schematically in Figure 15.2, photons at certain angles with the optical axis are used. One cone shown in the figure is ordinarily and the other extraordinarily polarized. The cones intersect along two directions where the light is unpolarized. Similarly to before, conservation laws and the coherence of the down-conversion process ensure that pairs of photons emerging along these intersections are entangled. Their

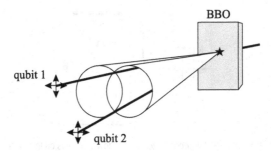

Fig. 15.2 Polarization-entangled photons generated by down-conversion in a BBO non-linear crystal. Photons traveling along the intersections between the two cones of ordinarily and extraordinarily polarized photons are in an entangled state.

state is given by

$$|\Psi\rangle = \frac{1}{\sqrt{2}} \left(|VH\rangle + |HV\rangle \right) \equiv \frac{1}{\sqrt{2}} \left(|10\rangle + |01\rangle \right).$$

The communication setups introduced below will mostly make use of polarization-entangled photons generated by BBO crystals.

15.2 Quantum dense coding

The idea behind quantum dense coding is that Alice and Bob can share a maximally entangled Bell state before Alice has decided which message to send to Bob. The communication scheme thus starts with, for example, the state $|\Psi^-\rangle$ shared between Alice and Bob. Alice then encodes one of four possible messages, 00, 01, 10, and 11, by applying one of four local quantum operations to her qubit of the entangled pair. This generates a different Bell state in each case, as shown in Table 15.1.[2] She then sends her qubit to Bob via the quantum channel and Bob measures both qubits in the Bell basis. Since the Bell states are orthogonal they can be distinguished using only one copy of the state. One of four messages, and therefore two bits of classical information, are transmitted from Alice to Bob in this scenario, while sending only one qubit from sender to receiver.

Table 15.1 Quantum dense coding with $|\Psi^-\rangle$.		
Message	Operation	Bell state
00	$\mathbb{1}$	$|\Psi^-\rangle$
01	σ_z	$|\Psi^+\rangle$
10	σ_x	$|\Phi^-\rangle$
11	$\sigma_x\sigma_z$	$|\Phi^+\rangle$

[2] This is always true only up to irrelevant global phases.

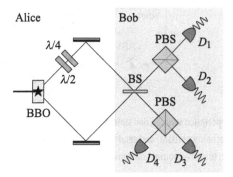

Fig. 15.3 Schematic experimental setup for quantum dense coding. A BBO crystal produces a polarization-entangled pair of photons. One photon is sent directly to Bob while Alice can use $\lambda/2$ and $\lambda/4$ plates to encode messages according to Table 15.1. Bob uses a partial Bell state analyzer to distinguish three different messages.

Note that at some point before the communication takes place Alice and Bob need to share the entangled pair. This requires Alice sending a qubit to Bob or the transmission of qubits to Alice and Bob from a source of entangled qubits. The entangled state is pure and independent of the message to be sent. The initial entangled pair thus acts as a resource for communication between Alice and Bob but is in itself not sufficient to send messages. This still requires transmission of one qubit *after* Alice has encoded her message in the Bell states.

15.2.1 Experimental setup

The schematic experimental setup first used for demonstrating quantum dense coding is shown in Figure 15.3. Parametric down-conversion creates a pair of polarization-entangled photons in a Bell state. One qubit is sent to the receiver and the other is manipulated by Alice before being submitted to Bob: a combination of $\lambda/4$ and $\lambda/2$ plates are used to realize the desired Bell state. When both qubits have arrived at the receiver side, Bob needs to carry out a Bell state measurement.

A setup for partially achieving a Bell state measurement is shown in Figure 15.3, which uses only standard optical elements.[3] Two polarization-entangled photons are incident onto the beam splitter (BS). The overall wavefunction of the two photons (polarization + spatial wavefunction) has to be symmetric, since the photons are bosons. Therefore, if the polarization part of the state is (anti)symmetric the spatial part also has to be (anti)symmetric. For the Bell state $|\Psi^-\rangle$, with antisymmetric polarization, one photon has to follow the upper arm and the other photon the lower arm after the beam splitter (see the following Example 15.1). The polarization of the photons is analyzed by sending them through a polarizing beam splitter (PBS) and then measuring horizontally and vertically polarized photons in separate photo-detectors D_1, D_2 for the upper arm and D_3, D_4 for the lower arm. Thus, a coincidence between D_3 and D_2 or D_4 and D_1 is registered for $|\Psi^-\rangle$. The remaining

---

[3] It turns out that all four Bell states can only be distinguished if non-linear optics is used.

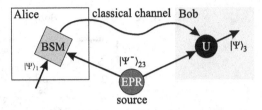

Fig. 15.4   Quantum teleportation setup. A Bell state is created at the source and distributed to Alice and Bob. Alice performs a Bell state measurement (BSM) and sends the result to Bob via a classical channel. Bob applies a unitary operation U conditional on the measurement outcome.

three Bell states have a symmetric polarization wavefunction and thus must also have a symmetric spatial wavefunction. Both photons follow the same arm for symmetric spatial wave functions. For $|\Psi^+\rangle$ a coincidence in D_1 and D_2 or in D_3 and D_4 is thus registered. In the other two cases $|\Phi^+\rangle$, $|\Phi^-\rangle$ two photons are detected in the same detector and these two Bell states can therefore not be distinguished. This analyzer identifies two of the four Bell states and distinguishes them from the other two. The setup thus allows the faithful transmission of three messages by sending one qubit from Alice to Bob.

Example 15.1 Two identical photons impinge on a 50/50 BS one arriving at the upper arm $|u\rangle$ and the other at the lower arm $|l\rangle$. The photons have an antisymmetric polarization wavefunction. Since the overall wavefunction is symmetric, the spatial part of the incoming wavefunction must be given by $(|ul\rangle - |lu\rangle)/\sqrt{2}$ for the overall wavefunction to be symmetric. Each photon undergoes a Hadamard gate in their spatial degrees of freedom at the beam splitter irrespective of polarization. This turns the wavefunction into

$$\frac{(|u\rangle + |l\rangle)(|u\rangle - |l\rangle) - (|u\rangle - |l\rangle)(|u\rangle + |l\rangle)}{\sqrt{8}} = \frac{|lu\rangle - |ul\rangle}{\sqrt{2}},$$

so the spatial part of the outgoing wavefunction is also antisymmetric. The photons thus follow different paths after the BS; the terms $|uu\rangle$ and $|ll\rangle$ cancel because of interference and so the photons will never travel along the same path. We note that this cancelation requires both photons to be indistinguishable. Perfect cancelation would not be possible for a state of the form $(|u\tilde{l}\rangle - |l\tilde{u}\rangle)/\sqrt{2}$ with different photon modes $|u\rangle \neq |\tilde{u}\rangle$ and $|l\rangle \neq |\tilde{l}\rangle$.

15.3 Quantum teleportation

The aim of quantum teleportation is to send an unknown quantum state of qubit 1 from Alice to Bob using classical communication and an entangled pair of qubits as shown in Figure 15.4. The sender Alice receives qubit 1 in state $|\Psi\rangle_1 = \alpha|0\rangle + \beta|1\rangle$ with unknown complex parameters α and β. She also receives qubit 2 which is part of an entangled Bell state $|\Psi^-\rangle_{23}$ with qubit 3 which is sent to Bob. The initial state of the three qubits

Table 15.2 Quantum teleportation with $|\Psi^-\rangle_{23}$		
BSM result	Projected state	Operation
$|\Psi^-\rangle_{12}$	$|\Psi^-\rangle_{12}(\alpha|0\rangle + \beta|1\rangle)_3$	$U = \mathbb{I}$
$|\Psi^+\rangle_{12}$	$|\Psi^+\rangle_{12}(\alpha|0\rangle - \beta|1\rangle)_3$	$U = \sigma_z$
$|\Phi^-\rangle_{12}$	$|\Phi^-\rangle_{12}(\alpha|1\rangle + \beta|0\rangle)_3$	$U = \sigma_x$
$|\Phi^+\rangle_{12}$	$|\Phi^+\rangle_{12}(\alpha|1\rangle - \beta|0\rangle)_3$	$U = \sigma_x\sigma_z$

is $|\Psi\rangle_{123} = [\alpha(|001\rangle - |010\rangle) + \beta(|101\rangle - |110\rangle)]$. Alice performs a Bell state measurement (BSM) on her two qubits and tells Bob the result via a classical channel. Each outcome is equally likely, so that two classical bits of information must be communicated between Alice and Bob. The measurement projects qubits 1 and 2 onto one of the four Bell states. The quantum state is projected according to the BSM measurement outcome as shown in Table 15.2. Note that qubit 3 is never entangled with qubits 1 or 2 after the BSM measurement.

Bob receives qubit 3 and the measurement result from Alice. He applies a unitary U to particle 3 conditional on the measurement result as shown in Table 15.2. For each measurement outcome the operation U turns the state of qubit 3 into the original state of qubit 1, yielding $|\Psi\rangle_3 = \alpha|0\rangle + \beta|1\rangle$ up to an irrelevant global phase. In Example 15.2 below a full quantum mechanical analysis of the protocol is provided.

The teleportation of one qubit starts with an entangled state of negative conditional entropy $S(\rho_A|\rho_B) = -1$, which is independent of the qubit state to be sent. The transmission of two bits of classical information from Alice to Bob via a classical channel is required to complete the protocol. Classical communication and entanglement are both essential resources for teleportation, but neither of them is sufficient on its own to teleport $|\Psi\rangle_1$ from Alice to Bob.

Example 15.2 The quantum state of all three qubits in the teleportation protocol before Alice's measurement can be written as

$$|\Psi\rangle_{123} = \frac{1}{2}\left(|\Psi^-\rangle_{12}(\alpha|0\rangle + \beta|1\rangle)_3 + |\Psi^+\rangle_{12}(\alpha|0\rangle - \beta|1\rangle)_3 + |\Phi^-\rangle_{12}(\alpha|1\rangle + \beta|0\rangle)_3 + |\Phi^+\rangle_{12}(\alpha|1\rangle - \beta|0\rangle)_3\right).$$

Thus each of the four Bell states will be found with probability 1/4. If the outcome is not revealed, the measurement turns the state into a mixed state given by

$$\rho_{123} = \left(|\Psi^-\rangle\langle\Psi^-|(\alpha|0\rangle + \beta|1\rangle)(\alpha^*\langle0| + \beta^*\langle1|) + |\Psi^+\rangle\langle\Psi^+|(\alpha|0\rangle - \beta|1\rangle)(\alpha^*\langle0| - \beta^*\langle1|)\right.$$
$$+ |\Phi^-\rangle\langle\Phi^-|(\alpha|1\rangle + \beta|0\rangle)(\alpha^*\langle1| + \beta^*\langle0|)$$
$$\left. + |\Phi^+\rangle\langle\Phi^+|(\alpha|1\rangle - \beta|0\rangle)(\alpha^*\langle1| - \beta^*\langle0|)\right)\frac{1}{4}.$$

At this stage the reduced density operator of Bob's particle is given by

$$\rho_3 = \text{Tr}_{12}\{\rho_{123}\} = \frac{1}{2}(|0\rangle\langle0| + |1\rangle\langle1|).$$

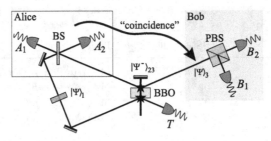

Fig. 15.5 Schematic experimental setup for quantum teleportation. Only the Bell state $|\Psi^-\rangle$ is identified in this experiment.

This is the maximally mixed state and has no correlations with the initial state of particle 1. Once the measurement outcome has been communicated from Alice to Bob, this additional information allows him to rule out three of the four terms in ρ_{123} and to apply the correct conditional unitary operation U in each case. We analyze its effect by studying what happens to the different parts of ρ_{123}, and find[4]

$$
\rho'_{123} = \frac{1}{4}\left(|\Psi^-\rangle\langle\Psi^-|+|\Psi^+\rangle\langle\Psi^+|+|\Phi^-\rangle\langle\Phi^-|+|\Phi^+\rangle\langle\Phi^+|\right)(\alpha|0\rangle+\beta|1\rangle)(\alpha^*\langle 0|+\beta^*\langle 1|)
$$

$$
= \frac{\mathbb{1}}{4}(\alpha|0\rangle+\beta|1\rangle)(\alpha^*\langle 0|+\beta^*\langle 1|).
$$

Alice possesses a maximally mixed state when averaging over many runs, but Alice and Bob of course know which Bell state is realized after each measurement. Alice's state is not correlated with Bob's particle and does not contain any information about α and β anymore. Bob is in possession of the state to be teleported. Note that it is required by the no-cloning theorem that the original qubit state is destroyed during the teleportation process.

15.3.1 Experimental setup

The schematics of the first experimental setup used for demonstrating teleportation with photons is shown in Figure 15.5. This protocol only works in a fraction of all runs and demonstrates the important tool of post-selection in quantum optics experiments.

A BBO crystal is used to create two entangled photon pairs. This is achieved by only keeping those runs where the strong light pulse entering the crystal produces a photon pair $|\Psi^-\rangle_{23}$, is then reflected at the mirror, and produces a second pair when traveling through the crystal again. The detector T is only used to trigger the experiment, it indicates that the second pair of photons has been created. Before arriving at Alice's site the second photon of this pair is brought into the desired initial state by a wave plate unknown to Alice. Alice then uses a minimal version of a Bell state analyzer which only detects the state $|\Psi^-\rangle_{12}$ through coincidence clicks in detectors A_1 and A_2. Any other event is disregarded.

---

[4] We do not write down every case individually but instead what happens on average assuming that the conditional operation is applied to the corresponding parts of the density operator.

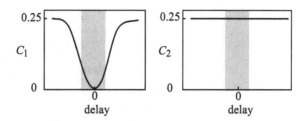

Fig. 15.6 Schematic experimental results for teleportation of a horizontally polarized photon. The two plots show coincidences C_1 between detectors A_1, A_2 and B_1 and C_2 between A_1, A_2 and B_2 as a function of the delay with which photons arrive at the BS. The shaded region indicates successful teleportation.

As shown in Table 15.2, Bob must apply the operation $\mathbb{1}$ for this outcome, that is do nothing. He then analyzes the polarization of his photon by using a PBS and two photo-detectors. This experiment applies the technique of post-selection since only those events are recorded where the detector T clicks and both of the detectors A_1 and A_2 click. Any other event is disregarded when analyzing the data.

A typical result of such an experiment is shown schematically in Figure 15.6 as a function of the delay between the arrival times of photons at detectors A_1 and A_2 for the teleportation of a horizontally polarized photon.[5] Coincidence probabilities C_1 between detectors A_1, A_2 and B_1, and C_2 between A_1, A_2 and B_2 are shown as a fraction of all events where two photon pairs were generated. When the delay between Alice's two photons is large, each of them has probability 1/2 of following either path after the BS, and there is thus an overall probability of 1/2 that they follow different paths after the BS. Teleportation is not successful in this case and thus both coincidences are equally likely, $C_1 = C_2 = 1/4$. When the delay tends toward zero the two photons will simultaneously arrive at the BS and interfere as discussed in Example 15.1. They will only follow different arms with probability 1/4, indicating the detection of the Bell state $|\Psi^-\rangle$.[6] Now teleportation is successful and $C_1 = 0$, while $C_2 = 1/4$. Therefore the dip in C_1 and the constant value of C_2 indicate successful teleportation of the horizontally polarized photon.

15.4 Entanglement swapping

The state of qubit 1 is teleported with all its quantum properties using the teleportation setup described in Section 15.3. In particular, any entanglement of qubit 1 with another system is preserved. We can see this by assuming α and β are state vectors instead of just c-numbers. This fact can be used to extend the teleportation setup to teleport entanglement. In this case the state of qubit 1 is not well defined but can always be written as $|\Psi\rangle_{14} = |\alpha\rangle_4 |0\rangle_1 + |\beta\rangle_4 |1\rangle_1$. The setup shown schematically in Figure 15.7 turns this state into the corresponding entangled state between qubits 3 and 4 given by $|\Psi\rangle_{34} = |\alpha\rangle_4 |0\rangle_3 + |\beta\rangle_4 |1\rangle_3$.

---

[5] Figure 15.5 does not show how to adjust this delay time.
[6] See also the Hong–Ou–Mandel effect in quantum optics.

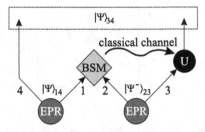

Fig. 15.7   Schematic entanglement swapping setup. The entanglement between qubits 1 and 4 created by the source is swapped to qubits 3 and 4 through teleporting qubit 1.

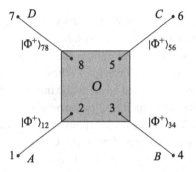

Fig. 15.8   A quantum telephone exchange.

No information about the original state will be left in qubits 1 and 2. The analysis of this setup using the same method as for the teleportation protocol is found in Exercise 15.9.

Example 15.3 We consider the quantum telephone exchange shown in Figure 15.8. Each user A, B, C, and D shares the Bell state $|\Phi^+\rangle$ with the central exchange as shown. Imagine that A, B and C would like to share the state $|\text{GHZ}\rangle = (|000\rangle + |111\rangle)/\sqrt{2}$ for further quantum communication.[7] The exchange O is asked to project particles 2, 3 and 5 into this GHZ state, which gives

$$_{235}\langle\text{GHZ}|(|\Phi^+\rangle_{12}|\Phi^+\rangle_{34}|\Phi^+\rangle_{56}) = (|000\rangle_{146} + |111\rangle_{146})/\sqrt{2}.$$

In this process the telephone exchange provides the communicating parties with an entangled state and becomes disentangled from all of them, and thus can not eavesdrop into their further communication. Furthermore, the initial preparation of the shared Bell pairs between users and O is independent of the states to be shared later between the users.

---

[7] GHZ states are discussed in the next chapter.

Further reading

A detailed description of the experimental setups discussed in this chapter can be found in Bouwmeester *et al.* (2000). This book also contains a description of optical techniques for quantum communication. The first experimental quantum teleportation was reported in Bouwmeester *et al.* (1997). A scalable scheme for quantum computing with linear optics was first introduced in Knill *et al.* (2001), and an accessible treatment of the Hong–Ou–Mandel effect can be found in Gerry and Knight (2005).

Exercises

15.1 Consider a momentum-entanglement interferometer experiment shown in Figure 15.1. Calculate the probability of coincidence clicks for all possible pairs of detectors assuming that the state

$$|\Psi\rangle = \frac{|ab\rangle + |ba\rangle}{\sqrt{2}}$$

is created in the down-conversion process as a function of the phase shift ϕ.

15.2 If down-conversion happened incoherently, photon pairs in state

$$|\Psi\rangle = \frac{1}{\sqrt{2}} \left(e^{i\varphi}|ab\rangle + |ba\rangle \right)$$

would be generated with a phase φ that takes on any value $[0, 2\pi]$ with equal probability. Show that in this scenario the photon pairs are described by the density operator

$$\rho = \frac{1}{2} \left(|ab\rangle \langle ab| + |ba\rangle \langle ba| \right).$$

15.3 Carry out the same calculation as in Exercise 15.1 but now for the state generated in an incoherent process as in Exercise 15.2.

15.4 Using the symmetry properties of photons, explain how the Bell states $|\Psi^{\pm}\rangle$ can be distinguished in a partial Bell state analyzer for polarization-encoded photons. Why can the other two states not be distinguished? Imagine that photons were fermions (which is not true!). How would the partial Bell state analyzer work in this case?

15.5 For quantum dense coding Bob needs a Bell state analyzer. What is the channel capacity (number of classical bits transmitted in one use of the channel) if Bob has an ideal Bell state analyzer? How is this channel capacity reduced if the Bell state analyzer is only able to identify the two Bell states $|\Psi^{\pm}\rangle$ but cannot differentiate between the two Bell states $|\Phi^{\pm}\rangle$?

15.6 Work out descriptions of the quantum dense coding and teleportation protocols for the case where the EPR source produces the Bell state $|\Phi^{+}\rangle$.

15.7 Show that without classical communication no information about the state of qubit 1 is transferred to qubit 3 in the teleportation protocol. Explain why it is impossible to transmit the quantum state of a qubit from Alice to Bob by classical communication only.

15.8 Alice and Bob carry out the teleportation protocol with an imperfect Bell state analyzer which cannot distinguish the states $|\Phi^+\rangle$ and $|\Phi^-\rangle$. Assume that Alice does not tell Bob about this imperfection but randomly assumes one of the two states whenever the Bell state analyzer gives an ambiguous result. Calculate the fidelity with which an arbitrary state $|\psi\rangle$ is teleported in this case. Which states are teleported with maximum fidelity and which states are teleported with minimum fidelity?

15.9 Qubit 1 is entangled with a quantum system 4. Their state can be written as

$$|\Psi\rangle_{14} = \frac{|0\rangle_1 \otimes |\phi\rangle_4 + |1\rangle_1 \otimes |\varphi\rangle_4}{\sqrt{2}} \, .$$

Show that if the teleportation protocol is applied to qubit 1 the entanglement with system 4 is swapped to qubit 3. Does this protocol work if qubits 1 and 4 are initially in a mixed entangled state?

16

Testing EPR

Quantum physics violates the basic assumptions of local realism. Particles can be entangled with each other in ways that allow correlations stronger than possible in classical physics. This was first realized by Einstein, Podolsky and Rosen (EPR) in 1935 and led them to conclude that quantum theory was incomplete. It took about 30 years before J. S. Bell found observable consequences of this phenomenon by showing how entangled states violate inequalities for two-particle correlations derived purely from the assumption that the world obeys local realism.

Violations of Bell inequalities were first observed experimentally in the Aspect experiment, which will be discussed below. Such experiments are technically challenging and care is needed to avoid any loopholes in the experimental setup that would still permit the description of the experiments by a local realistic theory.

Bell inequalities describe violations of local realism on average, meaning that each individual experimental outcome could occur in a local realistic theory. It is the statistics of these outcomes that violates local realism. In contrast, another class of quantum states, so-called Greenberger–Horne–Zeilinger (GHZ) states, which contain three particles, can be shown to violate local realism in each measurement outcome. We will also discuss this class of states and experiments by Zeilinger's group, which demonstrated this type of violation of local realism for the first time.

Our aim here is not to discuss the exact relations between quantum entanglement, non-locality and realism, and their physical and philosophical implications. While this is a highly interesting topic which attracts a lot of attention from researchers, such an exploration would go beyond the scope of this book.

16.1 Bell inequalities

Systems which follow classical common sense are described by a local and realistic theory, and can be shown to obey inequalities for the maximum strengths of correlations between their constituents. We will show that these inequalities are violated by quantum mechanics and discuss experiments demonstrating such violations.

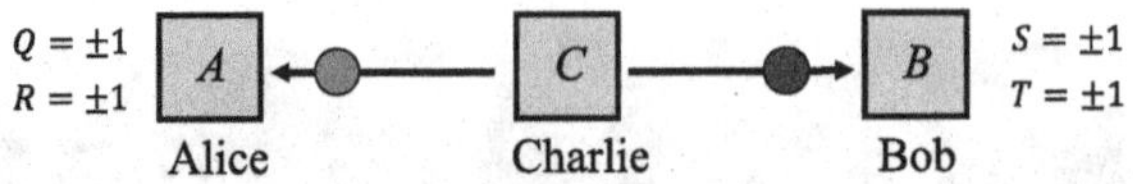

Fig. 16.1 Schematic setup to detect violations of the CHSH inequality.

16.1.1 The CHSH inequality

We derive a Bell-type inequality (the so-called CHSH inequality) by analyzing the Gedanken experiment shown in Figure 16.1 using common sense (not quantum theory). We assume:

- Charlie prepares two systems (possibly correlated) and sends one to Alice and the other one to Bob.
- After receiving their respective particles, Alice and Bob both randomly choose to measure one of two properties of their particle. Then they simultaneously perform their measurement.
- They repeat this experiment many times and record their outcomes.
- Alice and Bob meet and investigate the correlations between their experimental results.

What can they expect to obtain? For simplicity we assume that each measurement can only yield a value of ± 1. We describe the possible measurements of Alice by random variables Q and R and those of Bob by random variables S and T. By common sense we assume that the measurement values of Q, R, S, and T exist independent of observation, that is, whether or not the measurements needed to reveal them are actually made. This is the assumption of *realism*. Furthermore, Alice's measurement does not influence the outcome of Bob's measurement. They are performed in a causally disconnected manner, so it is reasonable to assume this. This is the assumption of *locality*.

We investigate the expression $QS + RS + RT - QT$, which can be written as

$$QS + RS + RT - QT = (Q + R)S + (R - Q)T.$$

Because we assume local realism, this can be simplified by noting that either $Q + R$ or $R - Q$ must be zero and the other combination must equal ± 2. Therefore

$$QS + RS + RT - QT = (Q + R)S + (R - Q)T = \pm 2.$$

Now we assume that the probability for measurement values $Q = q, R = r, S = s, T = t$ before the measurement is $p(q, r, s, t)$, and this exists and is not affected by the measurement because we assume local realism. Using this probability distribution and the linearity of expectation values we now work out the maximum value of the following Bell function $\mathcal{B}$ allowed by local realism:

$$\mathcal{B} = E(QS) + E(RS) + E(RT) - E(QT) = E(QS + RS + RT - QT),$$

where for instance $E(QS)$ denotes the expected value of products of simultaneous measurement outcomes of Q and S. We find

$$\mathcal{B} = E(QS+RS+RT-QT) = \sum_{q,r,s,t} p(q, r, s, t)(QS+RS+RT-QT) \leq 2 \sum_{q,r,s,t} p(q, r, s, t) = 2.$$

Fig. 16.2   Schematic experimental setup of the Aspect experiments.

This yields the so-called CHSH inequality obeyed if the assumptions of local realism hold for the measurements carried out by Alice and Bob, $\mathcal{B} \leq 2$.

We now analyze the same experiment using quantum mechanics for the case where Charlie generates an entangled Bell state $|\Psi^-\rangle$. He sends one qubit to Alice and the other to Bob. Alice chooses between measuring the operators $Q = \sigma_{1z}$ and $R = \sigma_{1x}$ on her qubit. Bob measures one of the operators $S = -(\sigma_{2z}+\sigma_{2x})/\sqrt{2}$ and $T = (\sigma_{2z} - \sigma_{2x})/\sqrt{2}$. We find the quantum mechanical expectation values $\langle QS \rangle = 1/\sqrt{2}$, $\langle RS \rangle = 1/\sqrt{2}$, $\langle RT \rangle = 1/\sqrt{2}$, and $\langle QT \rangle = -1/\sqrt{2}$ and therefore

$$\mathcal{B} = E(QS) + E(RS) + E(RT) - E(QT) = 2\sqrt{2} > 2 \,.$$

The experiment violates the CHSH inequality and therefore the assumptions of local realism. Entanglement between Alice's and Bob's states yields correlations stronger than allowed by local realism. Note that this violation is achievable only for some choices of measurements, and is maximal for the case discussed above.

Example 16.1 We work out which observables Q', R', S', and T' maximally violate the CHSH inequality for the Bell state $|\Phi^+\rangle$. We use the relation $|\Psi^-\rangle = \sigma_{1z}\sigma_{1x}|\Phi^+\rangle$ to write the expectation values for $|\Phi^+\rangle$ in terms of expectation values for $|\Psi^-\rangle$. For example,

$$\langle \Psi^-|QS|\Psi^- \rangle = \langle \Phi^+|\sigma_{1x}\sigma_{1z}Q\sigma_{1z}\sigma_{1x}S|\Phi^+ \rangle = \langle \Phi^+| - QS|\Phi^+ \rangle \equiv \langle \Phi^+|Q'S'|\Phi^+ \rangle \,.$$

By working out all four combinations of expected values appearing in the CHSH inequality we find that the observables $Q' = -Q$, $R' = -R$, $S' = S$, and $T' = T$ yield maximum violation of local realism for the Bell state $|\Phi^+\rangle$.

16.1.2 The Aspect experiments

The first experiments demonstrating the violation of Bell inequalities were carried out in the group of A. Aspect.[1] A schematic setup of the experiment is shown in Figure 16.2. The sources produce polarization-entangled photons in state $|\Psi^-\rangle$. One photon is detected by Alice after a polarizing beam splitter (PBS) at angle ϕ_A, the other at an angle ϕ_B by Bob. Setting Alice's PBS at angles $\phi_{A1} = 0$ and $\phi_{A2} = \pi/4$ corresponds to measuring observables Q and R, respectively. The angles $\phi_{B1} = \pi/8$ and $\phi_{B2} = 3\pi/8$ correspond to measuring observables T and S. Clicks in one of the two detectors correspond to

---

[1] We do not accurately describe these experiments but rather show schematically how their setup could be used to violate the CHSH inequality.

measurement outcomes ± 1, respectively. By correlating different measurement results, the Aspect experiments managed to demonstrate violation of the CHSH inequality.

Example 16.2 More generally, measuring at an angle ϕ with respect to the vertical polarization corresponds to projecting onto states $|\phi_+\rangle = \sin(\phi)|V\rangle + \cos(\phi)|H\rangle$ and $|\phi_-\rangle = \cos(\phi)|V\rangle - \sin(\phi)|H\rangle$. We assign measurement outcomes ± 1 to clicks in the corresponding detectors. The measurement then corresponds to measuring the operator $\sigma_\phi = |\phi_+\rangle \langle\phi_+| - |\phi_-\rangle \langle\phi_-| = \cos(2\phi)\sigma_z + \sin(2\phi)\sigma_x$.

16.1.3 Loopholes

The results of the Aspect experiments (and several experiments which followed them) can be viewed as evidence for the violation of local realism but this is not the only possible explanation. Various experiments have had several loopholes:

- They made a fair sampling assumption which presumes that the measured values are a fair reflection of all possible outcomes. This was necessary because of rather small photo-detector efficiencies which allowed measurement of only a small fraction of all events. This loophole can be closed by increasing the detector efficiency and nearly 100% detection efficiency has been achieved in ion trap experiments, but only at 3 μm separation of the Bell pair particles.
- Experimenters removed accidental coincidences from their data. By reducing imperfections in the experimental setups the number of such coincidences can be decreased. Then it becomes possible to keep such coincidences in the data without disturbing the measured correlations too much and still being able to violate Bell inequalities.
- In some experiments polarizers were set up before creation of the photon pairs and the procedure for setting them was not always guaranteed to be random. This allows for conspiracy theories which assume that the down-conversion process conspires with the polarizer setting to produce the "right kind" of photon pair for each setting. Such loopholes should be avoided: it is, for instance, possible to use a random quantum process to determine which measurement is to be carried out. For this the photon is sent through a beam splitter and measured differently depending on the path it takes.
- Strict Einstein locality was not obeyed in several experiments. This loophole has been closed in some experiments and measurements have even been carried out in different moving frames, finding violation of local realism and results consistent with quantum theory.
- In most experiments the quantum system under consideration is not truly a bipartite system. It consists, for example, of two photons and the atom which emitted them. In order to ensure Bell's inequality holds in this situation it must be shown that any auxiliary particles in the setup are not correlated with the two particles under consideration.

To date, most loopholes have been closed individually but no single experiment has closed all possible loopholes simultaneously.

16.2 GHZ states

Bell inequalities are formulated in terms of expectation values. In principle, the measurement of these expectation values requires infinitely many runs of the experiment. Local realism is only violated on average but not in any single run of an experiment. In contrast, quantum mechanics predicts the violation of local realism with certainty for GHZ states. In measurements on these states, outcomes which are forbidden according to local realism will be predicted by quantum theory to occur with finite probability. Since we can make definite predictions rather than statistical ones, no inequalities are needed in this setup.

A GHZ state is a three-qubit entangled state of the form

$$|\text{GHZ}\rangle = \frac{1}{\sqrt{2}} \left(|000\rangle + |111\rangle \right) = \frac{1}{\sqrt{2}} \left(|HHH\rangle + |VVV\rangle \right).$$

We analyze the state using notation commonly employed for polarization-encoded qubits; $|+\rangle \equiv |V'\rangle$ and $|-\rangle \equiv |H'\rangle$ are the polarizations rotated through $45°$ with respect to $|H\rangle$ and $|V\rangle$ and are eigenstates of σ_x. Left-handed $|L\rangle$ and right-handed $|R\rangle$ circular polarizations are eigenstates of σ_y. Rewriting the state $|\text{GHZ}\rangle$ in the YYX basis, we find

$$|\text{GHZ}\rangle = \frac{1}{2} \left(|RLH'\rangle + |LRH'\rangle + |LLV'\rangle + |RRV'\rangle \right).$$

Thus, if measuring in the YYX basis, we know with certainty the outcome of the third measurement after determining the state of the first two qubits. By cyclic permutation we find analogous expressions for measuring any two photons in circular polarization and the remaining one in $45°$ basis. In experiments, measurements are carried out in the three different bases XYY, YXY, and YYX to confirm these correlations. We derived them quantum mechanically here, but they are also allowed in local realistic theories. The difference appears when considering measurements in the XXX basis.

16.2.1 Local realistic analysis

From a local realistic point of view these perfect correlations can only be explained by assuming that each photon carries elements of reality which determine the outcome for all measurements in the XYY, YXY, and YYX bases. Which outcomes are possible in the XXX basis if these elements of reality exist? The previous measurements imply that if H' (V') is obtained for one photon, the other two have to have opposite (identical) circular polarizations. Imagine we find V' and V' for photons 2 and 3. Since 3 is V', 1 and 2 have to have identical circular polarization. Also, since 2 is V', 1 and 3 have to have identical circular polarization. All of these polarizations are elements of reality, so all photons have identical circular polarization. Thus, photon 1 needs to carry polarization V'. We conclude that $|V'V'V'\rangle$ is a possible outcome. Using similar arguments one can verify that under the assumption of local realism the only four possible outcomes are

$$|V'V'V'\rangle \qquad |H'H'V'\rangle \qquad |H'V'H'\rangle \qquad |V'H'H'\rangle.$$

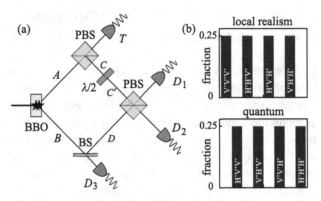

Fig. 16.3 (a) Schematic experimental setup of the GHZ experiment. Only events with two down-conversion processes are considered, the detector T acts as a trigger and detectors D_1, D_2 and D_3 are used to measure GHZ correlations. Linear optical elements for rotating the polarizations (not shown) can be used to conduct measurements in different bases. (b) Histogram of probabilities for different outcomes in the XXX basis as predicted by local realism and by quantum physics. Experimental results are consistent with quantum mechanics.

16.2.2 Quantum theoretical analysis

In the XXX basis the state $|\text{GHZ}\rangle$ reads

$$|\text{GHZ}\rangle = \frac{1}{2}\left(|H'H'H'\rangle + |H'V'V'\rangle + |V'H'V'\rangle + |V'V'H'\rangle\right).$$

The possible outcomes are therefore

$$|H'H'H'\rangle \qquad |H'V'V'\rangle \qquad |V'H'V'\rangle \qquad |V'V'H'\rangle.$$

The two sets of possible outcomes have no common elements, and thus any outcome that quantum mechanics predicts is not allowed by local realism and vice versa.

16.2.3 Experimental realization

The experimental setup is shown in Figure 16.3(a). Polarization-entangled pairs of photons in state

$$|\Psi\rangle = \frac{1}{\sqrt{2}}\left(|HV\rangle + e^{i\phi}|VH\rangle\right)$$

are created in a BBO crystal. Only the rare events where one UV pulse generates two such pairs are considered. As shown in Example 16.3 below, the resulting fourfold coincidence corresponds to the observation of the state

$$|\text{GHZ}\rangle = \frac{1}{\sqrt{2}}\left(|HHV\rangle + |VVH\rangle\right)$$

at the three detectors D_1, D_2, D_3 and the detection of the fourth photon at a detector T acts as a trigger. Note that in this experiment the coherence of the photons needs to be substantially longer than the length of the UV pulse, so that the two pairs created in the down-conversion processes are not distinguishable. Experimental results obtained from this setup agree with the quantum mechanical predictions shown in Figure 16.3(b) and are in disagreement with local realism.

Example 16.3 We analyze the setup shown in Figure 16.3 and show that coincidence clicks in all four detectors correspond to a GHZ state at detectors D_1, D_2 and D_3. We leave out normalization of the states for simplicity and will thus not be able to determine the fraction of events leading to coincidence clicks in all four detectors. The state created in the BBO is $|\psi\rangle_{tot} = (|H\rangle_A|V\rangle_B + e^{i\phi}|V\rangle_A|H\rangle_B)(|H\rangle_A|V\rangle_B + e^{i\phi}|V_A\rangle|H_B\rangle) = |H_A H_A V_B V_B\rangle + e^{i\phi}(|V_A H_A H_B V_B\rangle + |H_A V_A V_B H_B\rangle) + e^{i2\phi}|V_A V_A H_B H_B\rangle$. After the BS and the first PBS but before the $\lambda/2$ plate, the state is given by

$$|H_C H_C (V_D - V_3)(V_D - V_3)\rangle + \qquad \leftarrow \text{ no click in } T$$

$$2e^{i\phi}|V_T H_C (H_D - H_3)(V_D - V_3)\rangle + \qquad \leftarrow \text{ one click in } T$$

$$e^{i2\phi}|V_T V_T (H_D - H_3)(H_D - H_3)\rangle \qquad \leftarrow \text{ two clicks in } T.$$

We only keep those events where one click in T has occurred, and this part of the state is written as

$$|H_C H_D V_D\rangle + \qquad \leftarrow \text{ no click in } D_3$$

$$|H_C H_D V_3\rangle + |H_C V_D H_3\rangle + \qquad \leftarrow \text{ one click in } D_3$$

$$|H_C H_3 V_3\rangle \qquad \leftarrow \text{ two clicks in } D_3.$$

Again, only those events leading to one click in D_3 are kept. After the $\lambda/2$ plate, which is set up to implement a Hadamard gate, we obtain the state

$$|H_{C'} H_D V_3\rangle + |H_{C'} V_D H_3\rangle + |V_{C'} H_D V_3\rangle + |V_{C'} V_D H_3\rangle.$$

The second PBS transforms the state into

$$|H_1 H_2 V_3\rangle + |H_1 V_1 H_3\rangle + |V_2 H_2 V_3\rangle + |V_2 V_1 H_3\rangle.$$

Only the first and the last term yield exactly one click in each detector D_1, D_2 and D_3. Keeping only these events, we find

$$|\text{GHZ}'\rangle = |H_1 H_2 V_3\rangle + |V_1 V_2 H_3\rangle.$$

This state is locally equivalent to a standard GHZ state, that is, it can be turned into a GHZ state by applying a local unitary operation

$$\sigma_x^3|\text{GHZ}'\rangle = |\text{GHZ}\rangle = |HHH\rangle + |VVV\rangle.$$

Further reading

A comprehensive treatment and the history of research on Bell inequalities can be found in Bertlmann and Zeilinger (2002), which also discusses experiments that close some of the loopholes discussed in this chapter. More information on philosophical questions regarding this topic is contained in Mermin (1990). A detailed discussion of the GHZ experiment is contained in Bouwmeester *et al.* (2000). It is also interesting to read the original papers on the EPR paradox (Einstein *et al.*, 1935), Bell's derivation of his inequality (Bell, 1964), the Aspect experiments (Aspect *et al.*, 1981, 1982), GHZ states (Kafatos, 1989), and their first experimental verification (Pan *et al.*, 2000).

Exercises

16.1 Work out the quantum mechanical expectation values for the combination of observables Q, R at Alice's site and S, T at Bob's site for the setup described in Section 16.1.2. Show that they violate the CHSH inequality.

16.2 Derive a CHSH-type inequality which is violated if the EPR source produces the state $|\Phi^-\rangle$.

16.3 In the ZZZ basis a GHZ state is given by

$$|\text{GHZ}\rangle = \frac{1}{\sqrt{2}}(|HHH\rangle + |VVV\rangle).$$

By rewriting this GHZ state in the bases XYY, YXY and YYX, show that measuring two of the photons in circular polarization determines the polarization of the third photon in the X basis with certainty. Rewrite the GHZ state in the XXX basis and show that measuring in this XXX basis violates the expectations of local realism discussed in Section 16.2.1.

16.4 Calculate the probability with which the production of two photon pairs in the setup discussed in Section 16.2.3 leads to a click in all four detectors and hence work out the fraction of events which must be disregarded at the post-selection stage assuming that a single down-conversion process happens with a probability of 10^{-3} for each light pulse entering the BBO crystal.

17

Quantum cryptography

Cryptographic protocols can be classified by the type of security against eavesdropping which they provide. There exist mathematically secure schemes whose security relies on mathematical proofs (like the Vernam cipher discussed below) or conjectures[1] (like public key RSA encryption) about the complexity of deciphering the message without possessing the correct key. The majority of current secure public Internet connections rely on such schemes. Alternatively, a cryptographic setup may provide a physically secure method for communicating. In these setups the security is provided by the physical laws[2] governing the communication protocol. Here we first discuss a provably secure classical communication protocol and then quantum methods for distributing the necessary keys. The BB84 protocol relies on the impossibility of perfectly distinguishing non-orthogonal quantum states from one copy of the quantum system. The scheme was invented in 1984 and is the first of its kind. In contrast, the Ekert91 protocol makes use of Bell correlations between entangled pairs of photons. These correlations are destroyed when Eve attempts to make a measurement on one of the particles but also when imperfections, such as decoherence processes, affect the scheme. As long as no such "elements of reality" are introduced, and Bell correlations violating local realism can be generated by Alice and Bob, no eavesdropper can be present.

17.1   One-time pads and the Vernam cipher

The Vernam cipher is a cryptographic protocol which allows the encryption and decryption protocol to be publicly known. The security of the protocol relies entirely on the key which is private and not publicly known. Alice and Bob share identical n-bit secret key strings (one-time pad). Alice encodes her message by adding message and key using a classical XOR gate on each pair of bits. Bob decodes by subtracting the key again, that is, by applying another XOR operation with his key bit. As long as the key is of the same length as the message, is used only once, and can be securely distributed to Alice and Bob, the Vernam cipher is provably secure and Eve's mutual information with the sent message can be made arbitrarily small. Vernam ciphers require a secure method for distributing a large number of key bits, which must be delivered in advance of the message, otherwise one could deliver the message itself by secure means. Furthermore, the key bits must be guarded until they are used and the key must be destroyed after the bits are used. This key distribution problem

---

[1] This means that their security is not necessarily rigorously proven.
[2] But note that physical laws are never provably correct.

makes the Vernam cipher impractical for general use, but it is useful in diplomatic and espionage applications.

We note as an aside that the problem of key distribution is circumvented in public key cryptography. The public key can easily be used to encrypt a message, much like a box can be locked using a padlock without possessing the key. To decipher the message a private key – corresponding to the key for the padlock – needs to be used. In public key cryptography Alice sends out public keys to everyone and whoever wants to securely communicate with Alice may use her public key for encryption. The security of this protocol relies on the assumption that deciphering the message without possessing the private key – picking the padlock – is difficult.

17.2 The BB84 protocol

The BB84 protocol is a physically secure way to distribute a secret key with blocks of length n bits. It also allows detecting the presence of an eavesdropper Eve. Alice begins with two random bit strings A and B each consisting of $4n$ bits. She encodes these strings as a block of $4n$ qubits

$$|\psi\rangle = \bigotimes_{k=1}^{4n} |\psi_{a_k,b_k}\rangle,$$

where a_k is the kth bit of A and b_k is the kth bit of B. Each qubit is in one of the four states

$$
\begin{aligned}
|\psi_{00}\rangle &= |0\rangle, \\
|\psi_{10}\rangle &= |1\rangle, \\
|\psi_{01}\rangle &= |+\rangle, \\
|\psi_{11}\rangle &= |-\rangle.
\end{aligned}
$$

The bits in A are encoded in the basis X or Z as determined by B. These four states are not mutually orthogonal and cannot be distinguished with certainty. Alice sends the qubits to Bob who receives $\mathcal{E}(|\psi\rangle\langle\psi|)$, where $\mathcal{E}$ describes the action of the channel and an eventual eavesdropper. He publicly announces the fact that he has received the qubits. At this point Alice, Bob, and a possibly present Eve have their own states each with separate density matrices. Note that Alice has not revealed B, and so Eve has no knowledge on which basis she should have used when trying to eavesdrop the communication by measuring qubits. At best she can guess and if her guess is wrong she will disturb the states received by Bob. We note that noise in the channel also contributes to $\mathcal{E}$. Bob now measures each qubit in basis X or Z depending on a random $4n$-bit string B' which he creates on his own. We call Bob's measurement results A'. After this Alice announces B over a public channel and Bob and Alice discard all bits in $\{A, A'\}$ except for those where the bits in B and B' are equal. We assume that they can keep $2n$ bits.[3] It is important that Alice does not publish B

---

[3] In practice they will use strings of length $(4 + \delta)n$ with δ such that they are likely to obtain at least $2n$ cases where their bases agree.

before Bob has received the message to ensure security of the scheme! To check for noise and eavesdropping, Alice now selects n bits from A and publicly announces the selection. Alice and Bob publicly compare these n bits. If all of them agree, the probability of an eavesdropper being present is negligible and Alice and Bob share an n-bit secret key (the remaining bits of A).

In addition to the effects of eavesdropping, experimental imperfection will also lead to disagreement and this cannot be distinguished from an eavesdropper. If Alice and Bob find that t bits disagree, they can work out the maximum information that Eve could have gained and (i) abort and retry the protocol; (ii) use reconciliation and privacy amplification on the remaining n bits to obtain $m < n$ acceptably secret shared key bits; or (iii) decide that the achieved level of security is already sufficient and keep all n bits.

In the BB84 protocol, qubits need to be sent via a quantum channel and classical bits are also transmitted from Alice to Bob. However, it does not require any entanglement. The protocol relies on the fact that non-orthogonal quantum states cannot be perfectly distinguished by an eavesdropper whose actions will necessarily affect some of the qubits received by Bob. This reduces the mutual information between Alice and Bob for those cases where they have measured in the same basis. By detecting this reduction in mutual information they can identify Eve, as we will now investigate for a simple eavesdropping strategy.

17.2.1 Intercept–resend strategy

We investigate one particular eavesdropping strategy where Eve intercepts the sent qubits, measures them, and then resends them. This is called the intercept–resend strategy and we will see how Alice and Bob can detect Eve in this case and abort their communication.

Let us assume that Eve intercepts each qubit. She chooses the X or Z basis at random to measure the qubit. Then she prepares a qubit in the state she has measured and sends it to Bob. With 50% probability Eve will choose the wrong basis. Each time Eve's basis is wrong she will get a result which is completely uncorrelated with the bit that Alice has sent. If the channel is otherwise perfect, this leads to the outcomes and probabilities shown in Table 17.1 for the case where Alice sent message 0 encoded in basis X. Here $p(E|A)$ denotes the conditional probability for Eve's outcome given Alice's encoding and $p(B|EA)$ the conditional probability of Bob's outcome given Eve's outcome and Alice's encoding.

The cases where Alice sends 1X, 0Z, or 1Z can be worked out in the same way. Eve guesses the correct value of the bit with 75% probability. If Alice and Bob encode and measure in the same basis (i.e., in Table 17.1 considering only those cases where Bob measured in the X basis), then their results will disagree with a probability of 25%. For a perfect noiseless channel the mutual information between Alice's and Bob's messages obtained when measuring in the same basis has thus been reduced from $H(X : Y) = 1$ to $H(X : Y) = 0.189$ by Eve. The probability for Alice and Bob to find disagreement, thus identifying Eve, when comparing n of their key bits is given by

$$P_d = 1 - \left(\frac{3}{4}\right)^n .$$

Table 17.1 BB84: intercept–resend attack				
Alice	Eve	$p(E\|A)$	Bob	$p(B\|EA)$
0X	0X	1/2	0X	1/4
			0Z	1/8
			1Z	1/8
	1Z	1/4	1Z	1/8
			0X	1/16
			1X	1/16
	0Z	1/4	0Z	1/8
			0X	1/16
			1X	1/16

Conversely, if an eavesdropper intercepting every qubit should be detected with probability P_d, then the block length n must be chosen as

$$n = \frac{\log_2(1 - P_d)}{\log_2(3/4)}.$$

By sacrificing n bits from their key, Alice and Bob detect Eve with probability P_d if no noise is assumed. When the setup is imperfect, Alice and Bob need to tolerate a certain amount of disagreement and this will in general make it more difficult to spot an eavesdropper.

17.3 The Ekert91 protocol

An EPR source emits pairs of photons in the singlet state of polarizations

$$|\Psi^-\rangle = \frac{1}{\sqrt{2}}(|HV\rangle - |VH\rangle).$$

The two photons are distributed to Alice and Bob, respectively. They perform measurements and register the outcome by measuring at angles $\phi_{A1} = 0$, $\phi_{A2} = \pi/4$, $\phi_{A3} = \pi/8$ for Alice and by $\phi_{B1} = \pi/8$, $\phi_{B2} = 3\pi/8$, $\phi_{B3} = 0$ for Bob. The setup thus corresponds to the one shown in Figure 16.2, but Alice and Bob now choose between three different angles. Values of ± 1 are assigned to the measurement outcomes, as discussed in the Aspect experiments.

These angles are chosen independently and randomly for each pair. Alice and Bob measure a number of events and then publicly announce the orientations of the analyzers in each measurement. For the two pairs of angles (ϕ_{A1}, ϕ_{B3}) and (ϕ_{A3}, ϕ_{B1}), quantum mechanics predicts perfect anti-correlations between the measurement outcomes. Whenever Alice measures $+1$ then Bob has to necessarily measure -1. These measurement outcomes form the perfectly correlated secret key. Other measurements contain settings equivalent to what was used for violating Bell's inequality and are used to ensure security of the scheme. To do this Alice and Bob announce all results for which their orientations were different and they work out the value of $\mathcal{B}$ as in the Aspect experiments. This will only be $\mathcal{B} = 2\sqrt{2}$

if the particles were not disturbed and no eavesdropper was present. Furthermore, it ensures that measurements at angles (ϕ_{A1}, ϕ_{B3}) and (ϕ_{A3}, ϕ_{B1}) are indeed perfectly anti-correlated and form a reliable secret key. Eve cannot elicit any information from the particles while in transit from the source to the legitimate user since no information is encoded there. The information "comes into being" after the legitimate users perform the measurements and communicate in public afterwards. In each case an eavesdropper will introduce elements of physical reality to the particles, which will lower $\mathcal{B}$ below its quantum limit. Thus the Bell theorem can expose an eavesdropper.

Example 17.1 The setup in Figure 16.2 is often analyzed using correlation coefficients when studying quantum cryptographic schemes. The correlation coefficient of the measurements is given by

$$E(\phi_A, \phi_B) = P_{++}(\phi_A, \phi_B) + P_{--}(\phi_A, \phi_B) - P_{+-}(\phi_A, \phi_B) - P_{-+}(\phi_A, \phi_B),$$

where $P_{ab}(\phi_A, \phi_B)$ is the joint probability of Alice and Bob simultaneously measuring outcomes a and b, respectively when setting their detectors as angles ϕ_A and ϕ_B. Quantum mechanically the quantity $E(\phi_A, \phi_B)$ is therefore the expected value when measuring the operator $\sigma_{\phi_A} \otimes \sigma_{\phi_B}$. For a Bell state $|\Psi^-\rangle$ we find $E(\phi_A, \phi_B) = -\cos[2(\phi_A - \phi_B)]$ and the above results of perfect anti-correlation for settings (ϕ_{A1}, ϕ_{B3}) and (ϕ_{A3}, ϕ_{B1}), as well as violation of the CHSH inequality, follow.

In this scheme classical communication is only necessary to expose Eve, but not for establishing the secret key. If no eavesdropper could be present, all measurements could be carried out in the same basis. Before the measurement an entangled state with negative conditional entropy $S(\rho_A|\rho_B) = -1$ is created. The classical information gained from the measurements is perfectly correlated with $H(X : Y) = 1$. However, neither Alice nor Bob have the ability to engineer their measurement outcome and choose which message to send. Quantum mechanics does not tell us how to influence a measurement outcome and we do not have a more advanced theory to do this. Thus, while correlated classical information comes into being during the measurement process, it cannot be used to transmit messages between Alice and Bob. However, the random bits generated in this scheme are very useful as a secret key. The ideas introduced in the Ekert91 protocol lead on to device-independent quantum cryptography where security is based on the assumption of free will of Alice and Bob to make their measurement choices with the aim of being able to purchase secure communication devices from non-trustworthy sources.

17.4 Experimental setups

We can distinguish between free-space cryptography, where mostly polarization-encoded qubits are used, and fiber systems. Free-space experiments require robust optical setups at the sender and receiver side. Recent research on free-space cryptography has aimed to

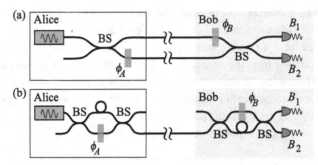

Fig. 17.1 BB84 using phase encoding in optical fiber setups. (a) Extended Mach–Zehnder setup. (b) Collapsed Mach–Zehnder setup. The circles denote delay loops of Δ and BS can be realized by bringing two fibers very close together.

increase the maximum distance over which single photons can faithfully be transmitted. Researchers achieved a distance of 144 km in an experiment carried out between La Palma and Tenerife in 2007. This marks an important step toward future satellite-based quantum communication and cryptography, with projects in this direction currently being pursued by a number of research groups. For instance, small integrated optical systems are currently being developed which will fit into cube-sats and could be launched into space as early as 2013. Here we consider in more detail two optical fiber setups to realize the BB84 protocol.

17.4.1 Phase-encoded fiber systems

Optical fibers do not conserve the polarization of photons because of randomly fluctuating birefringence (with a timescale of around one hour). Polarization tracking is possible but would make a polarization scheme cumbersome. Instead, we consider an extended Mach–Zehnder setup used for phase encoding and shown in Figure 17.1(a). Alice uses her phase modulator ϕ_A to encode 0, 1 in phases $\phi_{A1} = 0$ and $\phi_{A2} = \pi$ or in phases $\phi_{A3} = \pi/2$ and $\phi_{A4} = 3\pi/2$. Bob chooses between $\phi_{B1} = 0$ phase shift and $\phi_{B2} = \pi/2$ phase shift for his measurements. This scheme is equivalent to polarization encoding but replaces the polarization with a relative phase in the wavefunction. The drawback is that keeping the phase constant over large distances is very difficult due to temperature variations and other imperfections induced by the environment.

Example 17.2 We analyze the setup shown in Figure 17.1(a). The upper path corresponds to state $|0\rangle$ and the lower path to $|1\rangle$. Alice produces the state $|0\rangle$, which is turned into $|+\rangle$ by the BS and a relative phase ϕ_A is induced. The state which leaves the sender is $(|0\rangle + e^{i\phi_A}|1\rangle)/\sqrt{2}$. In the ideal case the delay Δ accumulated between Alice and Bob acts equally on both arms and gives $|0\rangle \rightarrow |0, \Delta\rangle$, and $|1\rangle \rightarrow |1, \Delta\rangle$. Since both arms are equally affected, we leave Δ out in the following. Bob introduces another relative phase ϕ_B so that the state turns into $(e^{i\phi_B}|0\rangle + e^{i\phi_A}|1\rangle)/\sqrt{2}$. The BS turns this state into $[(e^{i\phi_B} + e^{i\phi_A})|0\rangle + (e^{i\phi_B} - e^{i\phi_A})|1\rangle]/2$. If Alice chooses ϕ_{A1} (ϕ_{A2}) and Bob chooses ϕ_{B1}

he will get a click in B_1 (B_2) with certainty. If he chooses ϕ_{B2} a click in each detector is equally likely. If Alice chooses phases ϕ_{A3} (ϕ_{A4}) and Bob selects ϕ_{B2} B_1 (B_2) will click with certainty. Otherwise he will get equal probability for a click in one of the detectors. Thus Alice encodes 0 and 1 in the two bases by choosing from four different phases and Bob's measurements are carried out in the two different bases by choosing from two different phases. This realizes the BB84 protocol.

A more practical scheme is realized by collapsing the interferometer as shown in Figure 17.1(b). Two pulses are propagating down the same single fiber between Alice and Bob. The pulses are denoted by S (short path, no delay at the sender side) and L (long path, delay Δ at the sender side). The delay Δ is assumed to be much longer than the duration of the photon wave packet. After traveling through Bob's part of the Mach–Zehnder they create three different outputs. SS (which only experiences the delay between sender and receiver station) and LL (going through the delay lines at sender and receiver side in addition to the delay of the connecting fiber) are not relevant as they show no interference effects. SL and LS (going through exactly one of the delay lines at sender and receiver in addition to the delay of the fibre) are indistinguishable and thus interfere. Experimentally they can be selected by time gating. The choice of phase shifts by Alice and Bob gives the encoding–decoding in the SL and LS components exactly as in the previous scheme. This setup is more stable than that in Figure 17.1(a) since the pulses follow the same path for most of the setup. Any phase fluctuations which happen on timescales much longer than the delay Δ will only affect the global phase of the wavefunction, which is irrelevant. The major drawback of the scheme is that half of the signal is lost in the SS and LL path. The analysis of this scheme is similar to the analysis presented for the above setup, and left as Exercise 17.4. Note that the delay loops Δ at sender and receiver are present in one arm only. Their action can thus not be ignored in the analysis.

Further reading

An advanced description of the experimental setups discussed in this chapter can be found in Bouwmeester et al. (2000), which also contains a more detailed discussion of quantum eavesdropping strategies. The foundations of quantum cryptography were laid in Bennett and Brassard (1984) introducing the BB84 protocol and in Ekert (1991) proposing the Ekert91 scheme for quantum key distribution. The experiment on 144 km quantum communication is described in Ursin et al. (2007) and a review article covering quantum cryptography until 2002 is given in Gisin et al. (2002). Research projects integrating optical quantum cryptography systems with the aim of bringing them into space are, for instance, carried out by Ling (2011). Classical cryptography methods are described in detail in Schneier (1995).

Exercises

17.1 Assume that a communication channel used by Alice and Bob for BB84 key distribution is capable of transmitting 1000 qubits per second. What is the average key generation rate that Alice and Bob can achieve if (a) they assume that no eavesdropper can be present and thus do not publicly compare parts of their key; (b) an eavesdropper using intercept–resend strategy on each second qubit should be detected with 99.9% probability after two seconds? How much mutual information can be established between Alice's bit string A and the eavesdropper during these two seconds?

17.2 Calculate the joint probabilities $P_{ab}(\phi_A, \phi_B)$ introduced in Example 17.1 explicitly and use them to work out $E(\phi_A, \phi_B)$. Show that your result is consistent with directly calculating $E(\phi_A, \phi_B) = \langle \Psi^- | \sigma_{\phi_A} \otimes \sigma_{\phi_B} | \Psi^- \rangle$.

17.3 Show how a difference in the optical path length of the two fibers connecting Alice and Bob in Figure 17.1(a) leads to errors in the BB84 protocol. Assume that this phase error is equally distributed in the interval $\Delta\phi \in [-\pi/20, \pi/20]$. What is the probability that Alice and Bob obtain different measurement outcomes when publicly comparing bits measured in the same basis?

17.4 For the phase encoding systems in Figure 17.1(b) determine the probability for a photon to be incident on B_1 and B_2 as a function of the two phases induced by the two independent phase modulators ϕ_A and ϕ_B. Note that for the setup shown in this figure the photons going along paths SS and LL are assumed not to contribute to the signal. Explain how this setup can be used to realize the BB84 protocol.

Appendix: Quantum mechanics

Throughout this text we assume that the reader is familiar with elementary quantum mechanics and the properties of complex vector spaces, and in this appendix we provide a brief reminder of these topics. In particular, we introduce Dirac's notation for describing quantum mechanical systems. Many areas of quantum mechanics studied in undergraduate degrees can be described without using Dirac notation, and its importance is unclear. In other areas, however, the advantages of Dirac notation are huge, and it is essentially the only notation in use. This is particularly true of quantum information theory.

Dirac's notation is closely related to that used to describe abstract vector spaces known as Hilbert spaces, and many formal arguments about the properties of quantum systems are in fact arguments about the properties of Hilbert spaces. Here we aim to steer a careful course between the twin perils of excessive mathematical sophistication and of taking too much on trust. We will not prove some elementary results whose proof can be found elsewhere, but will concentrate on how these results can be used.

A.1 Hilbert space

A Hilbert space is an abstract vector space. As such, it has many properties in common with the use of ordinary three-dimensional vectors, but it also differs in several important ways. Firstly, the vector space is not three-dimensional, but can have any number of dimensions. (The description below largely assumes that the number of dimensions is finite, but it is also possible to extend these results to infinite-dimensional spaces.) Secondly, when the vectors are multiplied by scalar numbers these numbers can be complex. Thirdly, when two vectors are combined by taking their *scalar product* (analogous to the vector dot product, and often called the *inner product*), the result depends on the order in which the vectors are taken, such that

$$\mathbf{v}.\mathbf{u} = (\mathbf{u}.\mathbf{v})^* \tag{A.1}$$

where the asterisk indicates taking the complex conjugate. Clearly the scalar product of any vector with itself is real, as

$$\mathbf{u}.\mathbf{u} = (\mathbf{u}.\mathbf{u})^* \tag{A.2}$$

and the only numbers equal to their complex conjugates are real. It can also be shown that $\mathbf{u}.\mathbf{u}$ is positive, and its positive square root (the *norm* of $\mathbf{u}$) can be thought of as the length of $\mathbf{u}$.

As usual it is convenient to describe vectors by taking linear combinations of a set of basis vectors

$$\mathbf{v} = \sum_i \alpha_i \mathbf{u}_i, \tag{A.3}$$

where the α_i are complex coefficients and the $\mathbf{u}_i$ have the property that

$$\mathbf{u}_i . \mathbf{u}_j = \delta_{ij}, \tag{A.4}$$

where δ_{ij}, the Kronecker delta, is equal to 1 if $i = j$, and is equal to 0 if $i \neq j$. Such a basis is said to be *orthonormal*. The coefficients α_i can easily be found, as

$$\mathbf{u}_i . \mathbf{v} = \alpha_i \tag{A.5}$$

or

$$\mathbf{v} . \mathbf{u}_i = \alpha_i^* \tag{A.6}$$

where the second version follows from equation (A.1).

A.2 Dirac notation

The essence of Dirac notation is that the state of a quantum system is fully described by a vector in an associated Hilbert space. The notation makes a clear distinction between vectors appearing on the right-hand side and on the left-hand side of scalar products: vectors of the first kind are called *ket* vectors, or just kets, and are written as $|\psi\rangle$, while vectors of the second kind are called *bra* vectors, or bras, and written as $\langle\psi|$. The scalar product of a bra and a ket (usually called the *inner product*) is represented by the bra(c)ket notation

$$\langle\phi|\psi\rangle \tag{A.7}$$

and equation (A.1) is written as

$$\langle\phi|\psi\rangle = \langle\psi|\phi\rangle^*. \tag{A.8}$$

As before, bras and kets are conveniently expanded in an orthonormal basis, that is a set of kets such that

$$\langle i|j\rangle = \delta_{ij}. \tag{A.9}$$

Any ket $|\psi\rangle$ can then be written as

$$|\psi\rangle = \sum_i \alpha_i |i\rangle, \tag{A.10}$$

where

$$\langle i|\psi\rangle = \alpha_i. \tag{A.11}$$

The corresponding bra can be written as

$$\langle\psi| = \sum_i \alpha_i^* \langle i| \tag{A.12}$$

with

$$\langle\psi|i\rangle = \alpha_i^*, \tag{A.13}$$

so that the set of bras $\langle i|$ forms an orthonormal basis for the bras. The inner product between $\langle\phi|$ and $|\psi\rangle$ can now be written as

$$\langle\phi|\psi\rangle = \sum_i \sum_j \beta_i^* \langle i|\alpha_j|j\rangle = \sum_{i,j} \beta_i^* \alpha_j \langle i|j\rangle = \sum_{i,j} \beta_i^* \alpha_j \delta_{ij} = \sum_i \beta_i^* \alpha_i. \tag{A.14}$$

A.3 Operators

After kets and bras, the most important elements of Dirac notation are operators, which transform kets into other kets according to

$$A|\psi\rangle = |\psi'\rangle. \tag{A.15}$$

The action of an operator on a bra is analogous, but the operator must be written on the right-hand side of the bra:

$$\langle\phi|A = \langle\phi'|. \tag{A.16}$$

The relationship between these two actions is defined by the fact that

$$\langle\phi|\psi'\rangle = \langle\phi'|\psi\rangle \tag{A.17}$$

and so the inner product is written as

$$\langle\phi|A|\psi\rangle \tag{A.18}$$

and it is not necessary to specify whether the operator acts on the ket or the bra. These operators are linear, so that

$$A\left(|\psi\rangle + |\phi\rangle\right) = A|\psi\rangle + A|\phi\rangle \tag{A.19}$$

and

$$(A+B)|\psi\rangle = A|\psi\rangle + B|\psi\rangle. \tag{A.20}$$

The product of two operators acting on a ket is defined by acting first with the rightmost operator, so that

$$AB|\psi\rangle = A\left(B|\psi\rangle\right). \tag{A.21}$$

As discussed above, an operator can be thought to act either on a ket or on a bra, but these operations are not quite identical. In particular, the fact that $A|\psi\rangle = |\psi'\rangle$ does not in general imply that $\langle\psi|A = \langle\psi'|$. It is, however, true that

$$\langle\psi|A^\dagger = \langle\psi'|, \tag{A.22}$$

where $A^{\dagger}$ is an operator closely related to A, called the Hermitian conjugate or *adjoint* of A. The form of this operator will be considered below; for the moment it suffices to note that

$$\langle\phi|A|\psi\rangle = \langle\psi|A^{\dagger}|\phi\rangle^* \tag{A.23}$$

and this can be used to show that $(A^{\dagger})^{\dagger} = A$.

One important set of operators is the set of *projection operators*. Combining equations (A.10) and (A.11) gives

$$|\psi\rangle = \sum_i \langle i|\psi\rangle |i\rangle \tag{A.24}$$

and since the $\langle i|\psi\rangle$ inner products are just numbers, they can be swapped with the kets $|i\rangle$ to obtain

$$|\psi\rangle = \sum_i |i\rangle\langle i|\psi\rangle = \sum_i P_i|\psi\rangle, \tag{A.25}$$

where $P_i = |i\rangle\langle i|$ is an operator which projects $|\psi\rangle$ onto the basis ket $|i\rangle$, that is obtains the component of $|\psi\rangle$ which is parallel to $|i\rangle$. In the same way we can write

$$\langle\psi| = \sum_i \langle\psi|i\rangle\langle i| = \sum_i \langle\psi|P_i. \tag{A.26}$$

As the two equations above are valid for any ket or bra, it follows that

$$\sum_i |i\rangle\langle i| = \sum_i P_i = \mathbb{1} \tag{A.27}$$

where $\mathbb{1}$ is the *identity* operator, which leaves all bras, kets, and operators unchanged, so that

$$\mathbb{1}|\psi\rangle = |\psi\rangle \qquad \langle\psi|\mathbb{1} = \langle\psi| \qquad A\mathbb{1} = \mathbb{1}A = A. \tag{A.28}$$

This result is sometimes called the *closure theorem*.

Operators can be grouped into various classes according to their properties, and two particularly important groups are *Hermitian* and *unitary* operators. Hermitian operators are simply those which are equal to their adjoint:

$$H = H^{\dagger}, \tag{A.29}$$

while unitary operators have their inverse equal to their adjoint, so that

$$UU^{\dagger} = U^{\dagger}U = \mathbb{1}. \tag{A.30}$$

Most physical processes are described by Hermitian or unitary operators, and as we shall see below there is a close link between them.

A.4   Vectors and matrices

As shown in equation (A.10), any ket can be thought of as a linear combination of a set of orthonormal basis vectors. Provided there is some agreed basis, it clearly suffices just to list the coefficients: thus for a ket in a three-dimensional Hilbert space we can write

$$|\psi\rangle = \begin{pmatrix} \alpha_1 \\ \alpha_2 \\ \alpha_3 \end{pmatrix} \qquad (A.31)$$

where the coefficients form a column vector. A bra can be written in a similar way:

$$\langle\psi| = \begin{pmatrix} \alpha_1^* & \alpha_2^* & \alpha_3^* \end{pmatrix}, \qquad (A.32)$$

where the coefficients now form a row vector. Reconsidering equation (A.14) shows that when bras and kets are written in this form the inner product is nothing more than a conventional matrix product.

It is also possible to describe operators using a matrix. Clearly

$$A|\psi\rangle = \mathbb{1}A\mathbb{1}|\psi\rangle \qquad (A.33)$$

and applying the closure theorem gives

$$A|\psi\rangle = \sum_{i,j} |i\rangle\langle i|A|j\rangle\langle j|\psi\rangle \qquad (A.34)$$

$$= \sum_{i,j} \langle i|A|j\rangle\langle j|\psi\rangle|i\rangle, \qquad (A.35)$$

where we have used the fact that the two inner products in equation (A.34) are just numbers and so can be moved to the front of the formula. Next, note three things. Firstly, using equation (A.11), we know that $\langle j|\psi\rangle = \alpha_j$. Secondly, as $\langle i|A|j\rangle$ is just a number, we can choose to write it as an element A_{ij} of a matrix A. Finally, we can use equations (A.10) and (A.11) to expand $A|\psi\rangle$ in the same way as $|\psi\rangle$:

$$A|\psi\rangle = \sum_i \beta_i|i\rangle. \qquad (A.36)$$

Combining all these results gives

$$\beta_i = \sum_j A_{ij}\alpha_j \qquad (A.37)$$

and so the coefficients in the new state are obtained from those in the old state by multiplying them by A using conventional matrix multiplication.

Since a matrix can be used to describe an operator, it is instructive to consider how the product of two operators can be described. This can be achieved by considering a single element of the matrix description of the product

$$\langle i|BA|j\rangle = \langle i|B\mathbb{1}A|j\rangle \qquad (A.38)$$

$$= \sum_k \langle i|B|k\rangle\langle k|A|j\rangle \qquad (A.39)$$

or

$$(BA)_{ij} = \sum_k B_{ik}A_{kj} \qquad (A.40)$$

so that the matrix describing the product of two operators is simply the product of their individual matrices.

It is also instructive to consider the relationship between the matrix descriptions of an operator A and its adjoint $A^\dagger$. Applying equation (A.23) to the basis vectors gives

$$\langle i|A^\dagger|j\rangle = \langle j|A|i\rangle^* \tag{A.41}$$

or

$$(A^\dagger)_{ij} = A^*_{ji} \tag{A.42}$$

so that in matrix terms taking the adjoint is equivalent to taking the complex conjugate of the matrix transpose. From this fact it is straightforward to deduce that $(AB)^\dagger = B^\dagger A^\dagger$.

A.5 Eigenvalues and eigenvectors

Consider an operator A and a ket $|\psi\rangle$ such that

$$A|\psi\rangle = \lambda|\psi\rangle \tag{A.43}$$

where λ is just a number. The ket $|\psi\rangle$ is then said to be an eigenket of the operator A, with eigenvalue λ. Alternatively, and equivalently, the vector representation of $|\psi\rangle$ is an eigenvector of the matrix A with eigenvalue λ.

Eigenvalues are most conveniently determined using the matrix formalism. In a Hilbert space with n dimensions, equation (A.43) is equivalent to n simultaneous equations of the form

$$\sum_j A_{ij}a_j = \lambda a_i \tag{A.44}$$

or

$$\sum_j (A_{ij} - \lambda\delta_{ij})a_j = 0. \tag{A.45}$$

These simultaneous equations only have non-trivial solutions if the determinant of the coefficients on the left-hand side is zero, so that

$$\begin{vmatrix} (A_{11} - \lambda) & A_{12} & \dots & A_{1n} \\ A_{21} & (A_{22} - \lambda) & \dots & A_{2n} \\ \vdots & \vdots & \ddots & \vdots \\ A_{n1} & A_{n2} & \dots & (A_{nn} - \lambda) \end{vmatrix} = 0. \tag{A.46}$$

This determinant equation is in fact an nth-order polynomial in λ, whose n roots are the n eigenvalues of the matrix A. It should be noted that although the exact form of the determinant equation (A.46) seems to depend on the choice of basis in which A is described, the eigenvalues are fundamental properties of the operator A [equation (A.43)] and will be the same in any basis.

Once the eigenvalues have been determined, the eigenvector corresponding to each eigenvalue can be found by solving the set of simultaneous equations (A.45). Unlike eigenvalues, the eigenvectors of an operator obviously *do* depend on the basis used to describe the operator. The eigenkets of the operator, however, are fundamental properties and do not depend on the choice of basis.[1]

The method described above only gives the ratios of the coefficients describing the eigenvector, but this is quite proper as the eigenkets [equation (A.43)] are only defined up to a multiplicative factor. It is customary to choose kets of unit norm, but this does not completely define the ket, which can still be multiplied by any complex number of the form $e^{i\phi}$. A more important source of uncertainty, however, may arise when an operator has *degenerate* eigenvalues, arising from repeated roots in the eigenvalue polynomial. In this case linear combinations of eigenvectors corresponding to the same eigenvalue will also be eigenvectors with the same eigenvalue.

The process of finding eigenvalues and eigenvectors of a matrix is equivalent to *diagonalizing* the matrix: the matrix A can be written in the form

$$A = S\Lambda S^{-1} \tag{A.47}$$

where Λ is a diagonal matrix with the eigenvalues of A along the diagonal and S is formed from the eigenvectors of A. Equivalently, we can write any operator as a linear combination of its eigenkets $|\psi_i\rangle$, weighted by the corresponding eigenvalues λ_i,

$$A = \sum_i \lambda_i |\psi_i\rangle\langle\psi_i|. \tag{A.48}$$

Note that if we choose to work in the eigenbasis of A then this sum is obviously diagonal, and equal to Λ.

A.6 Operator trace

The trace of an operator is a particularly important property. As before, it is most simply defined by using a matrix description

$$\text{tr}(A) = \sum_i \langle i|A|i\rangle = \sum_i A_{ii} \tag{A.49}$$

but its value does not depend on the basis. This is most easily seen by writing the matrix A in diagonal form and then using the fact that the trace of a product of matrices is invariant under cyclic permutations of the product. Thus

$$\text{tr}(A) = \text{tr}(S\Lambda S^{-1}) = \text{tr}(\Lambda S^{-1}S) = \text{tr}(\Lambda) \tag{A.50}$$

and so the trace of an operator is equal to the sum of its eigenvalues.

---

[1] Here we assume that n linearly independent eigenvectors exist; see Section A.7 for more details.

A.7 Hermitian operators

As mentioned above, an operator A is Hermitian if it is equal to its adjoint, $A = A^\dagger$. Hermitian operators play a key role in quantum mechanics, and have many useful properties.

Firstly, the eigenvalues of a Hermitian operator are always real. Suppose $|a\rangle$ is an eigenket of A with eigenvalue a, so that

$$A|a\rangle = a|a\rangle, \tag{A.51}$$

or, equivalently,

$$\langle a|A|a\rangle = \langle a|a|a\rangle = a\langle a|a\rangle. \tag{A.52}$$

Using equation (A.23) gives

$$\langle a|A^\dagger|a\rangle = (\langle a|a|a\rangle)^* = a^*\langle a|a\rangle \tag{A.53}$$

and since $A = A^\dagger$ we can immediately deduce that $a = a^*$. Thus a must be real.

Secondly, the eigenkets of a Hermitian operator are mutually orthogonal. Consider two eigenkets such that

$$A|a_1\rangle = a_1|a_1\rangle \qquad A|a_2\rangle = a_2|a_2\rangle. \tag{A.54}$$

Since A is Hermitian, these can be rewritten as

$$\langle a_1|A = \langle a_1|a_1 \qquad \langle a_2|A = \langle a_2|a_2, \tag{A.55}$$

and so the inner product $\langle a_2|A|a_1\rangle$ can be expanded in two different ways:

$$\langle a_2|A|a_1\rangle = a_1\langle a_2|a_1\rangle = a_2\langle a_2|a_1\rangle \tag{A.56}$$

or

$$(a_1 - a_2)\langle a_2|a_1\rangle = 0. \tag{A.57}$$

The situation is simplest when the two eigenvalues are different, so that $(a_1 - a_2) \neq 0$; in this case equation (A.57) immediately requires that $\langle a_2|a_1\rangle = 0$, so that the kets $|a_2\rangle$ and $|a_1\rangle$ are orthogonal. Things are slightly more complex in the presence of degenerate eigenvalues, but in this case it can be shown that it is always possible to take linear combinations of the corresponding eigenkets to obtain orthogonal kets.

Taken together these results imply that for any Hermitian operator in an n-dimensional Hilbert space, it is always possible to find n orthonormal eigenkets of the operator. Clearly these orthonormal eigenkets provide a natural basis for describing the operator. This is not a formal proof, as it assumes that the eigenvalues and eigenkets always exist, but a more formal proof is possible using the *spectral decomposition theorem*, which states that any *normal* operator, that is any operator A for which $AA^\dagger = A^\dagger A$, is diagonal in some orthonormal basis. Clearly Hermitian matrices must be normal, and so the result is correct.

A.8 Commutators

When two operators, A and B, are applied in sequence to a ket $|\psi\rangle$ it usually matters which order they are applied in, so that

$$BA|\psi\rangle \neq AB|\psi\rangle \tag{A.58}$$

in general. More fundamentally, we note that operator multiplication (like matrix multiplication) is not *commutative*, so that $BA \neq AB$. In some cases, however, the operators do have the property that $BA = AB$, and in this case the operators are said to *commute*. This distinction is usually made by considering the commutator of the two operators

$$[A, B] = AB - BA \tag{A.59}$$

so that two operators commute if their commutator is zero. Note that for two operators to commute it must be true that $BA|\psi\rangle = AB|\psi\rangle$ for *every* ket $|\psi\rangle$, so that we can write $BA = AB$; it is not sufficient if the equality only holds for some particular kets.

Commutators play a key role in quantum mechanics, and it can be useful to consider their properties in the abstract. To give two trivial examples, it is obvious that

$$[B, A] = BA - AB = -[A, B] \tag{A.60}$$

and that

$$\text{tr}([A, B]) = \text{tr}(AB - BA) \tag{A.61}$$
$$= \text{tr}(AB) - \text{tr}(BA) \tag{A.62}$$
$$= 0, \tag{A.63}$$

where the last line has used the cyclic invariance of the trace.

A.9 Unitary operators

A unitary operator U was previously defined as an operator whose inverse is equal to its adjoint, but a more fundamental definition is that a unitary operator does not change the norm of a ket. We can now show how these two definitions are related. Consider some arbitrary ket $|\psi\rangle$, such that

$$U|\psi\rangle = |\psi'\rangle \quad \text{and} \quad \langle\psi|U^\dagger = \langle\psi'|. \tag{A.64}$$

It is clear that

$$\langle\psi'|\psi'\rangle = \langle\psi|U^\dagger U|\psi\rangle \tag{A.65}$$
$$= \langle\psi|U^{-1}U|\psi\rangle \tag{A.66}$$
$$= \langle\psi|\psi\rangle, \tag{A.67}$$

as required. In a similar way it can also be shown that unitary operators also leave the scalar product between any two kets unchanged. (This property suggests that unitary operators can be considered as changing between two different bases for describing a system, and this is indeed the case.)

As with Hermitian operators, unitary operators play a central role in quantum mechanics, and have many important features. For example, we note that the product of two unitary operators U and V is itself unitary, since

$$UV(UV)^\dagger = UVV^\dagger U^\dagger = UU^\dagger = \mathbb{1}. \tag{A.68}$$

More interestingly, it can be shown that the eigenvalues of a unitary operator all have modulus one, and that the eigenvectors of a unitary matrix are orthogonal. Both of these properties can be deduced by considering two eigenkets of U, $|u_1\rangle$ and $|u_2\rangle$, with eigenvalues λ_1 and λ_2. Clearly

$$\langle u_2|u_1\rangle = \langle u_2|U^\dagger U|u_1\rangle \tag{A.69}$$

$$= \lambda_2^*\lambda_1\langle u_2|u_1\rangle, \tag{A.70}$$

where the first line results from the fact that $U^\dagger U = \mathbb{1}$, and the second line comes from the fundamental properties of operators. Thus

$$(\lambda_2^*\lambda_1 - 1)\langle u_2|u_1\rangle = 0. \tag{A.71}$$

Choosing $|u_2\rangle = |u_1\rangle$ leads immediately to $\lambda_1^*\lambda_1 = 1$, showing that the eigenvalues have modulus one as required. Similarly, if $\lambda_1 \neq \lambda_2$, then the prefactor cannot equal zero, so the corresponding eigenvectors must be orthogonal. As before a more formal proof is possible using the spectral decomposition theorem, as unitary matrices are manifestly normal.

A.10 Operator exponentials

Next we consider an important link between unitary and Hermitian operators. Since the eigenvalues of a unitary operator have modulus one, they can all be written in the form

$$\lambda_j = \exp(-\mathrm{i}a_j) \tag{A.72}$$

where the numbers a_j are real. These numbers can be thought of as the eigenvalues of another operator A, which has the same eigenkets as U. Since the eigenvalues of A are real, A must itself be Hermitian. In general, we can write

$$U = \exp(-\mathrm{i}A) \tag{A.73}$$

connecting any unitary operator with its associated Hermitian operator.

The meaning of a function (such as the exponential) of an operator or matrix can easily be understood by considering a series expansion of the function. Thus

$$\exp(A) = \mathbb{1} + A + \tfrac{1}{2}AA + \ldots \tag{A.74}$$

and so on. Evaluating the series is simple for a diagonal matrix:

$$\exp\left[\begin{pmatrix} a & 0 \\ 0 & b \end{pmatrix}\right] = \begin{pmatrix} 1 & 0 \\ 0 & 1 \end{pmatrix} + \begin{pmatrix} a & 0 \\ 0 & b \end{pmatrix} + \frac{1}{2!}\begin{pmatrix} a & 0 \\ 0 & b \end{pmatrix}\begin{pmatrix} a & 0 \\ 0 & b \end{pmatrix} + \ldots \tag{A.75}$$

$$= \begin{pmatrix} 1 + a + a^2/2 + \ldots & 0 \\ 0 & 1 + b + b^2/2 + \ldots \end{pmatrix} \tag{A.76}$$

$$= \begin{pmatrix} \exp[a] & 0 \\ 0 & \exp[b] \end{pmatrix}. \tag{A.77}$$

The exponential of a general matrix can be calculated in a similar way by first diagonalizing the matrix and then noting that

$$\exp[S\Lambda S^{-1}] = S\exp[\Lambda]S^{-1}. \tag{A.78}$$

This result is easily proved by using a series expansion of the exponential function, as shown above, and canceling matching pairs of S^{-1} and S matrices. More fundamentally, S and S^{-1} are the matrices which transform between the basis we happen to be working in, and the *eigenbasis* of the operator, in which its description is naturally diagonal.

A.11 Analytical functions of operators

We can use the same method to calculate any analytical function of an operator, that is any operator which has a Taylor series expansion, and the same result can be derived more abstractly from the properties of operators. An operator can be written in terms of its eigenvalues and eigenvectors

$$A = \sum_i a_i |a_i\rangle\langle a_i|, \tag{A.79}$$

and the square can be written as

$$A^2 = \left(\sum_i a_i |a_i\rangle\langle a_i|\right)\left(\sum_j a_j |a_j\rangle\langle a_j|\right) = \sum_{ij} a_i a_j |a_i\rangle\langle a_i|a_j\rangle\langle a_j|. \tag{A.80}$$

Since the eigenvectors are orthonormal $\langle a_i|a_j\rangle = \delta_{ij}$, the sum can be simplified to

$$A^2 = \sum_i a_i^2 |a_i\rangle\langle a_i|. \tag{A.81}$$

In the same way we can write

$$A^n = \sum_i a_i^n |a_i\rangle\langle a_i|, \tag{A.82}$$

and by linearity any function of an operator can be written as

$$f(A) = \sum_i f(a_i)|a_i\rangle\langle a_i| \tag{A.83}$$

as long as f has a Taylor series expansion.

Note that in general it can be necessary to be careful with this definition, as some functions are only analytic for a restricted range of inputs. Throughout this text we only apply functions to operators where this can be done safely.

A.12 Physical systems

Now we can proceed to see how Dirac notation can be used to describe a physical system. The most important property of the system is its Hamiltonian operator $\mathcal{H}$, which describes the energy of the system. According to the time-independent Schrödinger equation, the Hamiltonian has an associated set of eigenstates

$$\mathcal{H}|j\rangle = \hbar\omega_j|j\rangle \qquad (A.84)$$

which form an orthonormal basis for the system. As the eigenvalues of the Hamiltonian (given by $\hbar\omega_j$) correspond to the energies of the eigenstates they must be real, and so $\mathcal{H}$ must be Hermitian. The most general state of the system is then some *superposition*, or linear combination, of these basis states:

$$|\psi\rangle = \sum_j \alpha_j|j\rangle. \qquad (A.85)$$

The evolution of the system is given by the time-dependent Schrödinger equation

$$\frac{\partial}{\partial t}|\psi\rangle = -\mathrm{i}\frac{\mathcal{H}}{\hbar}|\psi\rangle, \qquad (A.86)$$

which has the solution

$$|\psi(t)\rangle = U(t)|\psi(0)\rangle \qquad (A.87)$$

with

$$U(t) = \exp(-\mathrm{i}(\mathcal{H}/\hbar)t). \qquad (A.88)$$

The evolution of quantum states can also be described using the compact notation

$$|\psi\rangle \xrightarrow{\mathcal{H}t} U|\psi\rangle. \qquad (A.89)$$

Since $\mathcal{H}$ is Hermitian, the evolution operator U, usually called the *propagator*, must be unitary.

A.13 Time-dependent Hamiltonians

The discussion above assumes that the Hamiltonian is time-independent, that is it does not vary with time. This will not be true in complicated systems, which are controlled by varying the Hamiltonian. In many cases, however, the Hamiltonian is *piecewise constant*, that is it has a constant value for some finite length of time, and is then replaced by a different constant value for another finite time period, and so on. In this case the evolution can be described using a series of propagators

$$|\psi\rangle \xrightarrow{\mathcal{H}_1 t_1} \xrightarrow{\mathcal{H}_2 t_2} \xrightarrow{\mathcal{H}_3 t_3} U_3 U_2 U_1|\psi\rangle \qquad (A.90)$$

with $U_1 = \exp[-i(\mathcal{H}_1/\hbar)t_1]$ and so on. Note that the sequence of Hamiltonians is normally written with time running from left to right (that is the leftmost Hamiltonian is the first to be applied), while the sequence of propagators is always written from right to left, as the rightmost propagator is applied first. It is, of course, possible to combine the sequence of propagators into a single propagator, $U = U_3 U_2 U_1$, by matrix multiplication.

The situation is much more complicated when the Hamiltonian varies continuously with time. It is, of course, possible to write down a formal solution of the form of equation (A.90), but this is not generally a useful approach. A better approach is to remove the time-dependence of $\mathcal{H}$ by working in a time-dependent frame, and this approach is explored in Part I of the main text.

A.14   Global phases

The discussion above has glossed over one important aspect of using kets to represent the state of physical systems. The description of a physical state as a linear combination of basis states [equation (A.85)] provides *too much* information, as the kets $|\psi\rangle$ and

$$e^{i\phi}|\psi\rangle = \sum_j e^{i\phi}\alpha_j|j\rangle \tag{A.91}$$

describe the same physical state. It is safe to use this approach as long as you remember that two kets differing only by an overall phase shift correspond to the same state. Note also that states are only invariant under overall phases (often called *global phase shifts*); as we shall see, changes in the relative phases of the terms contributing to a superposition can be vitally important.

Further reading

While there are many undergraduate texts on quantum mechanics, most elementary texts avoid the use of Dirac notation, while many more advanced texts are largely concerned with applications of quantum mechanics in infinite-dimensional Hilbert spaces, such as the position and momentum representations. There are, however, several useful intermediate-level texts such as Gasiorowicz (2003), Bowman (2008), and Binney and Skinner (2010). Useful treatments can also be found in some textbooks on spin physics (e.g., Goldman, 1988).

Many of the properties of quantum mechanical systems are in reality just properties of the complex vector spaces used to describe them, and these are well described in both standard undergraduate texts, such as Riley *et al.* (2006), and in more advanced books such as Halmos (1974).

References

Abragam, A. 1983. *The Principles of Nuclear Magnetism*. Oxford: Oxford University Press.

Ashkin, A. 1997. Optical trapping and manipulation of neutral particles using lasers. *Proc. Natl. Acad. Sci. USA*, **94**, 4853–4860.

Aspect, A., Grangier, P., and Roger, G. 1981. Experimental tests of realistic local theories via Bell's theorem. *Phys. Rev. Lett.*, **47**, 460–463.

Aspect, A., Dalibard, J., and Roger, G. 1982. Experimental test of Bell's inequalities using time-varying analyzers. *Phys. Rev. Lett.*, **49**, 1804–1807.

Bakr, W. S., Peng, A., Tai, M. E., Ma, R., Simon, J., Gillen, J. I., Foelling, S., Pollet, L., and Greiner, M. 2010. Probing the superfluid to Mott insulator transition at the single atom level. *Science*, **329**, 547–550.

Beauchamp, K. G. 1987. *Transforms for Engineers*. Oxford: Clarendon Press.

Bell, J. S. 1964. On the Einstein–Podolsky–Rosen paradox. *Physics*, **1**, 195.

Bennett, C.H. 1973. Logical reversibility of computation. *IBM J. Res. Devel.*, **17**, 525–532.

Bennett, C.H. 1982. The thermodynamics of computation – a review. *Int. J. Theor. Phys.*, **21**, 905–940.

Bennett, C. H. and Brassard, G. 1984. Quantum cryptography: Public key distribution and coin tossing. Proc. IEEE Int. Conf. on Computers, Systems, and Signal Processing, Bangalore, p. 175.

Bennett, C. H. and DiVincenzo, D. P. 2000. Quantum information and computation. *Nature*, **404**, 247–255.

Bernstein, D. J., Buchmann, J., and Dahmen, E. (eds). 2010. *Post-Quantum Cryptography*. Berlin: Springer-Verlag.

Bertlmann, R. A. and Zeilinger, A. 2002. *Quantum [Un]speakables*. Berlin: Springer-Verlag.

Binney, J. and Skinner, D. 2010. *The Physics of Quantum Mechanics*. Great Malvern, UK: Capella Archive.

Blatt, R., Häffner, H., Roos, C. F., Cecher, C., and Schmidt-Kaler, F. 2004. Ion trap quantum computing with Ca^+ ions. *Quant. Inf. Proc.*, **3**, 61–73.

Bouwmeester, D., Pan, J-W., Mattle, K., Eibl, M., Weinfurter, H., and Zeilinger, A. 1997. Experimental quantum teleportation. *Nature*, **390**, 575.

Bouwmeester, D., Ekert, A., and Zeilinger, A. (eds). 2000. *The Physics of Quantum Information*. Berlin: Springer.

Bowman, G. E. 2008. *Essential Quantum Mechanics*. Oxford: Oxford University Press.

Braunstein, S. and Lo, H-K. 2000. Experimental proposals for quantum computation. *Fort. der Physik*, **48**, 767.

Budker, D., Kimball, D. F., and DeMille, D. P. 2004. *Atomic Physics*. Oxford: Oxford University Press.

Cirac, J. I. and Zoller, P. 1995. Quantum computations with cold trapped ions. *Phys. Rev. Lett.*, **74**, 4091–4094.

Cleve, R., Ekert, A., Macchiavello, C., and Mosca, M. 1998. Quantum algorithms revisited. *Proc. Roy. Soc. Lond. A*, **454**, 339–354.

Cohen-Tannoudji, C., Dupont-Roc, J., and Grynberg, G. 1992. *Atom–Photon Interactions*. Chichester: John Wiley & Sons.

Cory, D. G., Laflamme, R., Knill, E., Viola, L., Havel, T. F., Boulant, N., Boutis, G., *et al.* 2000. NMR based quantum information processing: Achievements and prospects. *Fort. der Physik*, **48**, 875–907.

Deutsch, D. 1985. Quantum theory, the Church–Turing principle and the universal quantum computer. *Proc. Roy. Soc. Lond. A*, **400**, 97–117.

DiVincenzo, D. P. 2000. The physical implementation of quantum computation. *Fort. der Physik*, **48**, 771–783.

Einstein, A., Podolsky, B., and Rosen, N. 1935. Can quantum-mechanical description of physical reality be considered complete? *Phys. Rev.*, **47**, 777.

Ekert, A. K. 1991. Quantum cryptography based on Bell's theorem. *Phys. Rev. Lett.*, **67**, 661.

Ernst, R. R., Bodenhausen, G., and Wokaun, A. 1987. *Principles of Nuclear Magnetic Resonance in One and Two Dimensions*. Oxford: Oxford University Press.

Estève, D., Raimond, J.-M., and Dalibard, J. (eds). 2003. *Quantum Entanglement and Information Processing*. Amsterdam: Elsevier.

Everitt, H. 2004. Special issue on experimental aspects of quantum computing. *Quant. Inf. Proc.*, **3**, 1–4.

Feynman, R. P. 1999. *Feynman Lectures on Computation*. London: Penguin Books.

Foot, C. J. 2005. *Atomic Physics*. Oxford: Oxford University Press.

Fredkin, E. and Toffoli, T. 1982. Conservative logic. *Int. J. Theor. Phys.*, **21**, 219–253.

Freeman, R. 1998. *Spin Choreography: Basic Steps in High Resolution NMR*. Oxford: Oxford University Press.

Gasiorowicz, S. 2003. *Quantum Physics*, 3rd edn. Chichester John Wiley & Sons.

Gerry, C. C., and Knight, P. L. 2005. *Introductory Quantum Optics*. Cambridge: Cambridge University Press.

Gisin, N., Ribordy, G., Tittel, W., and Zbinden, H. 2002. Quantum cryptography. *Rev. Mod. Phys.*, **74**(1), 145–195.

Goldman, M. 1988. *Quantum Description of High-Resolution NMR in Liquids*. Oxford: Oxford University Press.

Grover, L. K. 1997. Quantum mechanics helps in searching for a needle in a haystack. *Phys. Rev. Lett.*, **79**, 325–328.

Häffner, H., Roos, C. F., and Blatt, R. 2008. Quantum computing with trapped ions. *Phys. Rep.*, **469**, 155–203.

Halmos, P. R. 1974. *Finite-Dimensional Vector Spaces*. Berlin: Springer-Verlag.

Hecht, E. 2002. *Optics*, 4th edn. New York: Addison Wesley.

Hore, P. J. 1995. *Nuclear Magnetic Resonance*. Oxford: Oxford Chemistry Primers.

Hore, P. J., Jones, J. A., and Wimperis, S. 2000. *NMR: The Toolkit*. Oxford: Oxford Chemistry Primers.

Hughes, M. D., Lekitsch, B., Broersma, J. A., and Hensinger, W. K. 2011. Microfabricated ion traps. *Contemp. Phys.*, **52**, 505–529.

Hughes, R. 2004. Quantum information science and technology roadmap. Technical Report.

Jaksch, D., Briegel, H.-J., Cirac, J. I., Gardiner, C. W., and Zoller, P. 1999. Entanglement of atoms via cold controlled collisions. *Phys. Rev. Lett.*, **82**, 1975–1978.

Jessen, P. S., Deutsch, I. S., and Stock, R. 2004. Quantum information processing with trapped neutral atoms. *Quant. Inf. Proc.*, **3**, 91–103.

Jones, J. A. 2001. NMR quantum computation. *Prog. NMR Spectrosc.*, **38**, 325–360.

Jones, J. A. 2011. Quantum computing with NMR. *Prog. NMR Spectrosc.*, **59**, 91–120.

Kafatos, M. (ed.). 1989. *Bell's Theorem, Quantum Theory, and Conceptions of the Universe*. Dordrecht: Kluwer. Also available at http://arxiv.org/abs/0712.0921.

Knill, E., Laflamme, R., and Zurek, W. H. 1998. Resilient quantum computation. *Science*, **279**, 342.

Knill, E., Laflamme, R., and Milburn, G. J. 2001. A scheme for efficient quantum computation with linear optics. *Nature*, **409**, 46–52.

Ladd, T. D., Jelezko, F., Laflamme, R., Nakamura, Y., Monroe, C., and O'Brien, J. L. 2010. Quantum computers. *Nature*, **464**, 45–53.

Landauer, R. 1982. Uncertainty principle and minimal energy dissipation in the computer. *Int. J. Theor. Phys.*, **21**, 283–297.

Le Bellac, M. 2006. *Quantum Information and Quantum Computation*. Cambridge: Cambridge University Press.

Levitt, M. H. 2008. *Spin Dynamics: Basics of Nuclear Magnetic Resonance*, 2nd edn. Chichester: John Wiley & Sons.

Ling, A. 2011. See e.g. http://quantumlah.org/research/group/index.php?PI=21 and http://en.wikipedia.org/wiki/CubeSat.

Lipson, A., Lipson, S. G., and Lipson, H. 2011. *Optical Physics*, 4th edn. Cambridge: Cambridge University Press.

Mermin, N. D. 1990. *Boojums All The Way Through*. Cambridge: Cambridge University Press.

Mermin, N. D. 2007. *Quantum Computer Science*. Cambridge: Cambridge University Press.

Nielsen, M. A., and Chuang, I. L. 2000. *Quantum Computation and Quantum Information*. Cambridge: Cambridge University Press.

Ospelkaus, C., Warring, U., Colombe, Y., Brown, K. R., Amini, J. M., Leibfried, D., and Wineland, D. J. 2011. Microwave quantum logic gates for trapped ions. *Nature*, **476**(7359), 181–184.

Ozeri, R. 2011. The trapped-ion qubit tool box. *Contemp. Phys.*, **52**, 531–550.

Pan, J-W., Bouwmeester, D., Daniell, M., Weinfurter, H., and Zeilinger, A. 2000. Experimental test of quantum nonlocality in three-photon GHZ entanglement. *Nature*, **403**, 515.

Preskill, J. 1997–2011. Quantum computation. http://www.theory.caltech.edu/preskill/ph229/.

Riley, K. F., Hobson, M. P., and Bence, S. J. 2006. *Mathematical Methods for Physics and Engineering*, 3rd edn. Cambridge: Cambridge University Press.

Rueckner, W., Georgi, J., Goodale, D., Rosenberg, D., and Tavilla, D. 1995. Rotating saddle Paul trap. *Am. J. Phys.*, **63**, 186–187.

Ryan, C. A., Negrevergne, C., Laforest, M., Knill, E., and Laflamme, R. 2008. Liquid-state nuclear magnetic resonance as a testbed for developing quantum control methods. *Phys. Rev. A*, **78**, 012328.

Schneier, B. 1995. *Applied Cryptography: Protocols, Algorithms and Source Code in C.* Chichester: John Wiley & Sons.

Schumacher, B. and Westmoreland, M. D. 2010. *Quantum Processes, Systems, and Information.* Cambridge: Cambridge University Press.

Schweiger, A. and Jeschke, G. 2001. *Principles of Pulse Electron Paramagnetic Resonance.* Oxford: Oxford University Press.

Sherson, J. F., Weitenberg, C., Endres, M., Cheneau, M., Bloch, I., and Kuhr, S. 2010. Single-atom-resolved fluorescence imaging of an atomic Mott insulator. *Nature*, **467**, 68–72.

Shor, P. W. 1999. Polynomial-time algorithms for prime factorization and discrete logarithms on a quantum computer. *SIAM Rev.*, **41**, 303–332.

Slichter, C. P. 1989. *Principles of Magnetic Resonance.* Berlin: Springer-Verlag.

Southwell, K. 2008. Quantum coherence. *Nature*, **453**, 1003.

Stolze, J. and Suter, D. 2008. *Quantum Computing*, 2nd edn. New York: Wiley-VCH.

Suter, D. and Mahesh, T. S. 2008. Spins as qubits: Quantum information processing by nuclear magnetic resonance. *J. Chem. Phys.*, **128**, 052206.

Timoney, N., Baumgart, I., Johanning, M., Varon, A. F., Plenio, M. B., Retzker, A., and Wunderlich, Ch. 2011. Quantum gates and memory using microwave-dressed states. *Nature*, **476**(7359), 185–188.

Ursin, R., Tiefenbacher, F., Schmitt-Manderbach, T., Weier, H., Scheidl, T., Lindenthal, M., Blauensteiner, B., *et al.* 2007. Entanglement-based quantum communication over 144 km. *Nature Physics*, **3**, 481.

Vandersypen, L. M. K. and Chuang, I. L. 2004. NMR techniques for quantum control and computation. *Rev. Mod. Phys.*, **76**, 1037–1069.

Vedral, V. 2005. *Modern Foundations of Quantum Optics.* London: Imperial College Press.

Vedral, V. 2006. *Introduction to Quantum Information Science.* Oxford: Oxford University Press.

Weitenberg, C., Endres, M., Sherson, J. F., Cheneau, M., Schausz, P., Fukuhara, T., Bloch, I., and Kuhr, S. 2011. Single-spin addressing in an atomic Mott insulator. *Nature*, **471**, 319–324.

Wiesner, S. 1983. Conjugate coding. *SIGACT News*, **15**, 78–88.

Wineland, D. J. and Leibfried, D. 2011. Quantum information processing and metrology with trapped ions. *Laser Phys. Lett.*, **8**, 175–188.

Wooters, W. K. and Zurek, W. H. 1982. A single quantum cannot be cloned. *Nature*, **299**, 802–803.

Index

Printed in the United States
By Bookmasters